1991

The Federal Communications Commission

ELECTRONIC MEDIA GUIDES

The Federal Communications Commission

A Primer

Robert L. Hilliard

Emerson College

Focal Press
Boston   London

Focal Press is an imprint of Butterworth–Heinemann.

Recognizing the importance of preserving what has been written, it is the policy of Butterworth–Heinemann to have the books it publishes printed on acid-free paper, and we exert our best efforts to that end.

Library of Congress Cataloging-in-Publication Data
Hilliard, Robert L.,
  The Federal Communications Commission: a primer / Robert L. Hilliard.
          p.   cm. — (Electronic media guides)
  Includes index.
  ISBN 0-240-80101-6 (paperback)
  1. United States. Federal Communications Commission.
  2. Telecommunication policy—United States. I. Title. II. Series.
  Electronic media guide.
  HE7781.H54 1991
  353.0087'4—dc20
                                                    90-44813
                                                    CIP

British Library Cataloguing in Publication Data
Hilliard, Robert L.
  The Federal Communications Commission: a primer-
  (Electronic media guides)
  1. United States. Broadcasting services
  I. Title II. Series
  353.0087454

  ISBN 0-240-80101-6

Butterworth–Heinemann
80 Montvale Avenue
Stoneham, MA 02180

10   9   8   7   6   5   4   3   2   1

Printed in the United States of America

*To Lee and John for their courage, humor, dedication, caring, and love of life —
and to all the others who also put the quality of the individual above the quantity of
the bureaucracy.*

Contents

Preface

This book is based principally on official public documents of the Federal Communications Commission and other government agencies and offices, and on the author's almost two decades in Washington, D.C., as Chief of the FCC's Educational/Public Broadcasting Branch, chair of the Federal Interagency Media Committee, and in other federal communications and education positions.

The Federal Communications Commission: A Primer is designed to provide to both the communications student and practitioner a basic introduction to the mission, jurisdiction, organization, and operations of the FCC, including its licensing and rule-making procedures, method and process of implementing the Communications Act of 1934, historical foundations, and relationships to other government entities, the communications industry, and the general public. It includes descriptions of the responsibilities and work of the commissioners and all of the Commission's offices.

One purpose of this book is to fill a void in communications study. While few courses are devoted entirely to the FCC, many courses devote several or more sessions to the FCC. It is hoped that this primer will be the much-needed, useful text for such study. And, while not eliminating their need for professional communication attorneys and engineers, it is also hoped that this book will provide broadcasters, common carriers, private radio users, and others who have to deal with the FCC with some insights and understanding of how they can most simply and efficiently meet FCC requirements.

I am grateful to Mike Keith and Phil Sutherland for their help in generating the idea for this book; to Allen Myers, Bill Harris, the Consumer Assistance and Small Business Division of the Office of Public Affairs; and others at the FCC for providing significant information and material; and to Focal Press senior editor Karen Speerstra.

Robert L. Hilliard
Cambridge, Massachusetts
January 1991

1

What Is the FCC?

The Federal Communications Commission (FCC) is an independent government agency responsible directly to Congress. Established by the Communications Act of 1934, it is charged with regulating interstate and international communications by radio, television, wire, satellite, and cable. Its jurisdiction covers the 50 states and territories, the District of Columbia, and United States possessions.

The degree and nature of regulation is determined by the *Rules and Regulations* promulgated by the FCC, and varies with the philosophy of the President of the United States and the party in power at any given time. For example, the pro-consumer attitudes of the John F. Kennedy appointees to the FCC resulted in strong regulation in the public interest for many years, while the marketplace theories of the Ronald R. Reagan appointees resulted in deregulatory policies that eliminated many FCC regulations and diminished others.

In addition, congressional legislation and federal court decisions may add to or uphold, or reduce or abolish FCC responsibilities and authority. For example, the Cable Communications Act of 1984 eliminated almost all of the FCC's regulatory powers over cable, and on two occasions in the 1980s the federal courts found the FCC's cable must-carry requirement of local television stations unconstitutional.

THE COMMUNICATIONS ACT OF 1934

Section I of the Act describes its purposes in creating the Federal Communications Commission:

> For the purpose of regulating interstate and foreign commerce in communication
> by wire and radio so as to make available, so far as possible, to all the people of
> the United States a rapid, efficient, Nation-wide, and world-wide wire and radio
> communication service with adequate facilities at reasonable charges, for the
> purpose of the national defense, for the purpose of promoting safety of life and
> property through the use of wire and radio communication, and for the purpose
> of securing a more effective execution of this policy by centralizing authority
> heretofore granted by law to several agencies and by granting additional author
> ity with respect to interstate and foreign commerce in wire and radio communi
> cation, there is hereby created a commission to be known as the "Federal
> Communications Commission," which shall be constituted as hereinafter pro
> vided, and which shall execute and enforce the provisions of this Act.

1 ▲

Section 2 of the Act defines the FCC's jurisdiction as applying to

... all interstate and foreign communications by wire or radio and all interstate and foreign transmission of energy by radio, which originates and/or is received within the United States, and to all persons engaged within the United States in such communication or such transmission of energy by radio, and to the licensing and regulating of all radio stations as hereinafter provided.

Section 3 of the Act defines the means of communication and other key terms under FCC jurisdiction. Some of the principal definitions are

(a) "Wire communication" or "communication by wire" means the transmission of writing, signs, signals, pictures, and sounds of all kinds by aid of wire, cable, or other like connection between the points of origin and reception of such transmission, including all instrumentalities, facilities, apparatus, and services ... incidental to such transmission.

(b) "Radio communication" or "communication by radio" means the transmission by radio of writing, signs, signals, pictures, and sounds of all kinds, including all instrumentalities, facilities, apparatus, and services ... incidental to such transmission.

(c) "Licensee" means the holder of a radio station license granted or continued in force under the authority of this Act.

(d) "Transmission of energy by radio" or "radio transmission of energy" includes both such transmission and all instrumentalities, facilities, and services incidental to such transmission.

(e) "Interstate communication" or "interstate transmission" means communication or transmission (1) from any State, Territory, or possession of the United States, or the District of Columbia, to any other State, Territory, or possession of the United States, or the District of Columbia ... between points within the United States but through a foreign country ... but shall not ... include wire or radio communications between points in the same State, Territory, or possession of the United States, or the District of Columbia, through any place outside thereof, if such communication is regulated by a State commission.

(f) "Foreign communication" or "foreign transmission" means communication or transmission from or to any place in the United States to or from a foreign country, or between a station in the United States and a mobile station located outside the United States.

(h) "Common carrier" or "carrier" means any person engaged as a common carrier for hire, in interstate or foreign communication by wire or radio or in interstate or foreign radio transmission of energy ... a person engaged in radio broadcasting shall not, insofar as such person is so engaged, be deemed a common carrier.

(k) "Radio station" or "station" means a station equipped to engage in radio communication or radio transmission of energy.

(o) "Broadcasting" means the dissemination of radio communications intended to be received by the public, directly or by the intermediary of relay stations.

(p) "Chain broadcasting" means simultaneous broadcasting of an identical program by two or more connected stations.

(q) "Amateur station" means a radio station operated by a duly authorized person interested in radio technique solely with a personal aim and without pecuniary interest.

The Act's definitions further include "telephone," "radio telegraph," and a number of items relating to "ship" or "vessel" radio communication. The General Provisions under Title I also describe the organization and functioning of the Commission and its staff.

Title II describes the FCC's jurisdiction over common carriers, including service and charges. Title III covers radio, including allocation of facilities, licensing of stations, equal time facilities for candidates for public office (Section 315), operations of transmitters, broadcast matter and content, radio operations and requirements on ships, and provisions of the Public Broadcasting Act of 1967 that pertain to facilities grants and the establishment of the Corporation for Public Broadcasting and national systems of public television and public radio.

Title IV delineates the administrative procedures of the Commission. Title V deals with penal provisions for violations of the Act, and includes detailed prohibitions on rigging contests—reaction to the infamous quiz-show scandals. Title VI covers miscellaneous provisions such as unauthorized interception of communications, presidential powers over communications during a war emergency, and provision of telephone service for the disabled.

The Communications Act of 1934, as amended, also contains several appendices. Appendix A relates to the administrative procedures and judicial review of the FCC, with such items as public information, rule-making procedures, case hearings, and imposition of sanctions. Appendix B has definitions of federal agencies' relationships to the federal courts. Appendix C, constituting only one page in the Act, nevertheless contains provisions that have had widespread effects: broadcasting lottery information, using communications for fraud, and broadcasting "obscene, indecent, or profane language." Appendix D is the "Communications Satellite Act of 1962," which created a communications satellite corporation, COMSAT.

Some laws dealing with communications are enacted as entities unto themselves and not solely as amendments to the Communications Act. These laws are incorporated into FCC policy by the nature of their relationship to FCC responsibilities. The Communications Act is affected, as well, by noncommunications legislation that, in one or more of its provisions, relates to matters under FCC jurisdiction.

An important example of such legislation is the Subject Matter and Scope of Copyright Act of 1976, which not only revised U.S. copyright law, but included provisions affecting secondary transmissions in broadcasting and cable. It established a compulsory license for cable with a system of fees and a Copyright Royalty Tribunal to administer the fee collection and distribution. This compulsory license, giving individual cable systems access to copyright materials for one comprehensive fee, continues to be an important bargaining chip in the broadcasting-cable rivalry. Noncommunications legislation such as the Coordination of Federal Information Policy Act of 1980 and the Paperwork Reduction Omnibus Act of the same year required the FCC to modify its policies and practices in these two areas relative to its requirements of licensees and its liaison with the public.

The Cable Communications Act of 1984 virtually eliminated FCC jurisdiction over cable. The Consolidated Budget Reconciliation Act of 1985 established fee schedules for FCC application, license, and other services (see Chapter 5). The Comprehensive Smokeless Tobacco Health Act of 1986 included an extension of the FCC's responsibilities regarding the banning of advertising of cigarettes on the electronic media to smokeless tobacco. The Electronic Communications Privacy Act of 1986 charged the FCC with enforcing a prohibition on the interception of wire, electronic, or oral communication in FCC-regulated services.

Other Communications Act duties of the FCC affected by recent legislation came through the Lotteries Mail Fraud Act of 1988, which extended the ban on the use of broadcasting for such fraud to cable and subscription television (STV), and the Telecommunications Accessibility Enhancement Act of 1988, which concerns authorization of a Telecommunications Device for the Deaf (TDD)

FCC RULES AND REGULATIONS

The duties and jurisdiction of the FCC, as authorized in the Communications Act of 1934, as amended, are implemented through the FCC's Rules and Regulations, a body of specific definitions, requirements, and interpretations that have been developed by the FCC over its many years of existence. The Rules and Regulations are contained in what is called Title 47 of the *Code of Federal Regulations* (CFR), and consists of five separate volumes, obtainable only from the Superintendent of Documents (see Chapter 6). Each volume covers a general area of FCC responsibility, and is divided into parts covering the services regulated, as follows:

Volume I (Parts 0–19)

      Part 0—Commission Organization

      Part 1—Practice and Procedure

      Part 2—Frequency Allocation and Radio Treaty Matters

      Part 5—Experimental Radio Services

      Part 13—Commercial Radio Operators

      Part 15—Radio Frequency Devices

      Part 17—Construction, Marking, and Lighting of
            Antenna Structures

      Part 18—Industrial Scientific and Medical Equipment

      Part 19—Employee Responsibilities and Conduct

Volume II (Parts 20–39)

      Part 21—Domestic Public Fixed Radio Services

      Part 22—Public Mobile Service

      Part 23—International Fixed Public Radiocommunication
            Services

      Part 25—Satellite Communications

      Part 32—Uniform System of Accounts for
            Telecommunications Companies

      Part 34—Uniform System of Accounts for Radiotelegraph
            Carriers

Part 35—Uniform System of Accounts for Wire-telegraph
  and Ocean-cable Carriers
Part 36—Jurisdictional Separation Procedures; Standard
  Procedures for Separating Telecommunications
  Property Costs, Revenues, Expenses, Taxes, and
  Reserves for Telecommunications Companies

Volume III (Parts 40–69)

Part 41—Telegraph and Telephone Franks
Part 42—Preservation of Records of Communication
  Common Carriers
Part 43—Reports of Communication Common Carriers and
  Certain Affiliates
Part 61—Tariffs
Part 62—Applications to Hold Interlocking Directorates
Part 63—Extension of Lines and Discontinuance of Services
  by Carriers and Grants of Recognized Private
  Operating Agency Status
Part 64—Miscellaneous Rules Relating to Common Carriers
Part 65—Interstate Rate of Return Prescription
  Procedures and Methodologies
Part 66—Applications Relating to Consolidation,
  Acquisition, or Control of Telephone Companies
Part 68—Connection of Terminal Equipment to the
  Telephone Network
Part 69—Access Charges

Volume IV (Parts 70–79)

Part 73—Radio Broadcast Services
Part 74—Experimental, Auxiliary, and Special Broadcast
  and Other Program Distribution Services
Part 76—Cable Television Service
Part 78—Cable Television Relay Service

Volume V (Parts 80–100)

Part 80—Stations in the Maritime Service
Part 87—Aviation Services
Part 90—Private Land Mobile Radio Services
Part 94—Private Operational Fixed Microwave Services
Part 95—Personal Radio Services
Part 97—Amateur Radio Service
Part 99—Disaster Communication Service
Part 100—Direct Broadcast Satellite Service

Chapter 5 of this book explains the process by which FCC Rules and Regulations are enacted.

FCC STATUS

The Federal Communications Commission is one of the independent agencies of the U. S. government. Although falling under the Executive Branch, the FCC is a *regulatory* rather than an *executive* agency. While independent executive agencies and regulatory agencies both are headed by administrators appointed by the President and confirmed by the Senate, the former serve at the pleasure of the President, while the latter serve for fixed terms and cannot be removed by the President. Regulatory agencies, in addition, usually are composed of Commissioners, Directors, or Governors, with one political party (the party in power at the time) permitted no more than a majority of one.

Relationship to the Executive Branch

While the Chair of the FCC (the FCC still officially uses the term Chairman, which will, accordingly, be used in this book) and the Commissioners are nominated by the President, they do not report to the White House. Nevertheless, all appointees to any position owe a political debt to the person who appointed them. The President ostensibly does not get involved in the operations of the independent regulatory agencies. However, the chair of a given agency is, with rare exceptions, a member of the President's party, and is likely to come in contact with the President during his or her tenure, even if only at ordinary social or political events. One notable departure from the Executive Branch's direct continuing relationship to an FCC Chairman through party politics was Democrat Lyndon B. Johnson's naming of sitting Republican Commissioner Rosel H. Hyde as Acting Chairman, then Chairman following the departure of Democrat Chairman E. William Henry in 1966. One explanation is that President Johnson wished to avoid the appearance of directly influencing the FCC in light of his personal extensive communications holdings.

Relationship to Congress

Control and oversight of the executive departments are functions of Congress in its authority to enact laws, appropriate funds, and make rules for government operations. In that sense, the FCC is directly responsible to Congress. In the Senate the FCC falls under the jurisdiction of the Committee on Commerce, Science, and Transportation, and more specifically under the Subcommittee on Communications. In the House of Representatives the Committee on Energy and Commerce is the overseeing body, with the Subcommittee on Telecommunications, Consumer Protection, and Finance taking direct responsibility. In both the Senate and the House independent agencies are subject, as well, to the decisions of the respective Committees on Appropriations. In the Senate the FCC falls under the Subcommittee on Commerce, Justice, State, the Judiciary, and Related Agencies, and in the House under the Subcommittee on Commerce, Justice, State, and Judiciary.

The appropriate committee or subcommittee may at any time hold a hearing on an amendment to the Communications Act of 1934, summon the FCC for a briefing, require written submissions from the FCC, require the FCC to testify at a bill consideration, and otherwise maintain continuing check on the FCC's operations.

Quite obviously, control of the purse strings is the ultimate power. When the Federal Trade Commission (FTC) defied Congress in 1979 and decided to go ahead with its proposed investigation into the harm the contents of breakfast cereals may cause to children's health, Congress simply did not renew its appropriation for the coming year, thus putting the FTC out of business. The FTC canceled its investigation.

The FCC, as do other agencies, prepares its budget requests in the summer. These are reviewed in detail in the fall by the Office of Management and Budget (OMB) and presented to the president in the context of the overall budget plan. The funding recommendations are made by the president to Congress, which begins its formal review in January. Two steps are taken, first the enacting of legislation authorizing the outlay, and second legislation legally appropriating the money. Individual agencies are responsible for administering their budgets in accordance with the specified approved allocations. OMB reviews programs and financial reports and the General Accounting Office (GAO) does a regular audit.

Because each Commissioner nominated by the President must be confirmed by the Senate, appointees must have sufficient political connections and support to be appointed. It is rare that a commissioner is approved by the Senate without the approval of both of the senators from his or her home state. As in the relationship to the President, any appointee not only must be politically acceptable to a majority of the Senate, but owes a debt to the senators who promoted his or her candidacy.

Relationship to Other Federal Agencies

As an independent regulatory agency, the FCC is not obligated to work or consult with other federal agencies. But as a practical matter it finds it necessary to do so because of complementary and sometimes overlapping jurisdiction of specific communication matters. For example, the National Telecommunications and Information Administration (NTIA) in the Department of Commerce has responsibility for the future development of telecommunications and information technology and services; the FCC maintains continuing liaison with the NTIA. When planning for World Administrative Radio Council (WARC) meetings on the use of international frequency space, the FCC works closely with the State Department, which also represents the United States at such meetings. At one time, when this author was chair of the Federal Interagency Media Committee, an organization of some 30 federal agencies seeking the most efficient use and development of federal telecommunications efforts through cooperative projects, the FCC participated in the combined work of the member agencies as it related to FCC concerns. The FCC frequently finds itself relating to other federal agencies, sometimes in a cooperative, sometimes in an adversarial role, ranging from the Department of Justice (on such items as minority preferences and other legal issues) to the Federal Trade Commission (for example, false or misleading advertising on television or radio) to the Department of Interior (special frequency assignments for Native American reservations or Alaskan outposts).

SPECTRUM MANAGEMENT

The Federal Radio Commission (FRC), the predecessor of the FCC, was established in 1927 for the principal purpose of managing the spectrum. Although the Department of Commerce had been licensing radio stations since 1921, it had no jurisdiction over frequency or power assignments, and radio stations went on the air at will, creating literal chaos on the airwaves. The Radio Act of 1927 gave the newly created FRC authority to assign frequencies, lessening interference potentials. In 1934 the creation of the FCC added expanding telecommunications services such as television, satellite, and microwave to the government's spectrum management responsibilities. Many people in the communication industries oppose what they believe is too strong regulation in other than technical areas and maintain that spectrum management should still be the FCC's principle, if not only, authorized responsibility.

The FCC and NTIA are the two federal agencies responsible for allocating and managing the spectrum. The FCC has jurisdiction over all nonfederal government operations, and NTIA oversees federal government operations, including that of the military.

Frequency Assignments

While most of us are familiar with radio and television broadcasting frequencies, many other services have frequency allocations, including aeronautical and maritime communications and navigation; land mobile communications for business, industrial, public safety, and transportation services; amateur radio and personal communications, such as *ham* operators; microwave relay, satellite communications, telephone, data, closed-circuit television; and others. Current technology is able to make use of the spectrum space between 9 *kilohertz* (kHz—one thousand cycles or waves per second) and 40 *gigihertz* (GHz—one million kilohertz). While that may seem like a lot of space, as the demand for services in existing technologies grows and new technologies are developed, the available spectrum space has become severely limited.

The spectrum is divided into different band groups, each band allocated to specified kinds of service. For example, between 10 and 540 kHz is allocated primarily to long-range radiotelegraph and beacons for ships and aircraft; the 106 frequencies between 540 kHz and 1600 kHz (being expanded to 1705 kHz in the 1990s) are reserved for AM broadcasting; 1605 kHz to 25 *megahertz* (MHz—one thousand kHZ) is reserved for long-distance radiotelegraph and telephone, and for ships, planes, and international broadcasting; FM and TV have individual allocations between 25 and 890 megahertz; above that are bands allocated to radio navigation, common carrier, mobile, and many other specialized services. Experimental services are sometimes authorized in the spectrum above 40 GHz.

Criteria The criteria used by the FCC for frequency allocations is based on public need and benefit, technical considerations, and apparatus limitations.

Under public considerations the FCC determines whether the service needs radio spectrum rather than using wire lines, the probable number of people who will benefit from the service, the social and economic importance of the service—including safety of life and property, the degree of public support the service is likely to receive, whether the service should be made available on an extended or limited basis, the geographic area to be served, and the impact of the new service on existing investments in the frequency band.

Under technical considerations the FCC judges the amount of spectrum needed as a whole and for each channel, the propagation characteristics of the service and their compatibility with other services in and out of the frequency band, the signal strength and potential interference from and to the new service, and whether the proposed technology is proven and available or still under development.

Under apparatus limitations the FCC evaluates the operating characteristics of the transmitters, including the upper practical limit of the useful frequency spectrum, their power and the ability to stay on frequency, the practical limitations of the antennas to be used, and the availability and usefulness of the receiver for the service.

Frequency assignments are made on a primary and secondary basis. Primary allocations have priority in using the assigned frequency. Secondary assignments may operate as long as they don't cause interference to a primary assignment; when they do, they must cease operation or make modifications that remove interference factors.

International Assignments   International uses of the spectrum are coordinated by the International Telecommunications Union (ITU), of which some 165 countries are members. The ITU publishes the *International Radio Regulations*, which includes allocations and technical rules for the three regions covering the world. The allocations are developed at meetings of the World Administrative Radio Conference (WARC), which meets every 10 years and has frequent regional and special meetings in between. The FCC and NTIA, plus the Department of State, represent the United States at these meetings.

2

▼
▼
▼
▼
▼

Organization and Operation: Commissioners and Staff Offices

The FCC is directed by five commissioners, with the Chairman, designated by the President, as chief executive officer. Reporting directly to the Chairman and the Commissioners are the offices of the Managing Director, Plans and Policy, the Review Board, Administrative Law Judges, Congressional and Public Affairs, Engineering and Technology, and the General Counsel. Four Bureaus, also reporting directly to the Commission, conduct the principle FCC operations. These are the Mass Media, Common Carrier, Private Radio, and Field Operations Bureaus. Within each Bureau are Divisions and Branches designed to carry out responsibilities in designated areas. (see Figure 1).

THE COMMISSIONERS

The five FCC Commissioners serve 5-year terms. The Chairman delegates management and administrative responsibilities to the Managing Director, and other functions to staff units and bureaus and to committees of the Commissioners. The Commission—the body of Commissioners making up the FCC—holds regular meetings, some open and some closed to the public, and special meetings called by the chairman, to consider items prepared by the staff for a weekly agenda of business. Sometimes special items are brought to the attention of the Commission by one or more of the Commissioners. On occasion the Commission does its business by *circulation* —a procedure by which a given document or agenda item is submitted to each Commissioner for consideration and official approval or rejection outside of the Commission meeting.

The Chairman presides over all FCC meetings, coordinates and organizes the work of the Commission, and represents the Commission in legislative matters and in relations with other government agencies. If the Chairman is absent or the office is vacant and the President has not appointed one of the sitting Commissioners to the post, the Commissioners designate one of their members to serve as Acting Chairman.

Commissioner Staffs

With dozens of agenda items each week, ranging through all the varied communications services and involving legal, technical, and economic issues, it is impossible for a Commissioner to handle all the work alone. Each Commissioner is

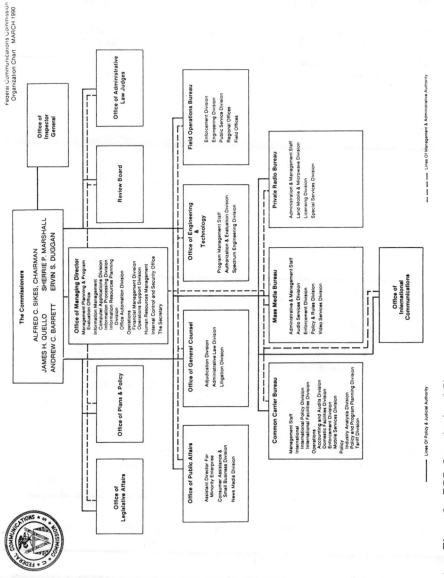

Federal Communications Commission
Organization Chart MARCH 1990

The Commissioners
ALFRED C. SIKES, CHAIRMAN
JAMES H. QUELLO SHERRIE P. MARSHALL
ANDREW C. BARRETT ERVIN S. DUGGAN

Office of
Inspector
General

Office of Administrative
Law Judges

Review Board

Office of Managing Director
Management Planning & Program
Evaluation Office
Information Management
Computer Applications Division
Information Processing Division
Information Resources Planning
Division
Office Automation Division
Operations
Financial Management Division
Operations Support Division
Human Resources Management
Internal Control and Security Office
The Secretary

Field Operations Bureau
Enforcement Division
Engineering Division
Public Service Division
Regional Offices
Field Offices

**Office of Engineering
&
Technology**
Program Management Staff
Authorization & Evaluation Division
Spectrum Engineering Division

Office of Plans & Policy

Office of General Counsel
Adjudication Division
Administrative Law Division
Litigation Division

Private Radio Bureau
Administration & Management Staff
Land Mobile & Microwave Division
Licensing Division
Special Services Division

Mass Media Bureau
Administrative & Management Staff
Audio Services Division
Enforcement Division
Policy & Rules Division
Video Services Division

Office of
Legislative Affairs

Office of Public Affairs
Assistant Director For
Minority Enterprise
Consumer Assistance &
Small Business Division
News Media Division

Common Carrier Bureau
Management Staff
International
International Policy Division
International Facilities Division
Operations
Accounting and Audits Division
Domestic Facilities Division
Enforcement Division
Mobile Services Division
Policy
Industry Analysis Division
Policy and Program Planning Division
Tariff Division

Office of
International
Communications

——— Lines Of Policy & Judicial Authority

– – – – Lines Of Management & Administrative Authority

▲ *Figure 1 FCC Organizational Chart.*

provided with funds for a professional and service staff, which can be used in any way the Commissioner sees fit. Usually, a Commissioner will have a legal assistant and an engineering assistant. Sometimes a Commissioner who happens to be a communications lawyer or an engineer may elect to forego an assistant in that area and employ someone in a different specialization, such as economics—an important aspect of FCC work. Some Commissioners have special, confidential assistants, and all have secretarial staffs.

Commission Meetings

The Commissioners' assistants receive and evaluate the agenda items supplied by the staff of the Bureau or Office preparing the item, usually at least a week before the commission meeting, and brief their respective commissioners prior to the meeting. At the Commission meetings the bureau or office chief and whomever the chief designates—frequently the head of the division or branch and the individual staff member or members who prepared the item—make a brief presentation to the commissioners. The Commissioners sit *en banc*, similar to judges in a courtroom, and question the presenters of the item, then have an open discussion among themselves before voting to accept, reject, or send the item back for further work.

If the item requires evaluation or eventual enforcement by additional offices of the Commission, representatives from the other offices may also participate in the discussion. For example, a Mass Media Bureau agenda item concerning a cable must-carry rule, which would have to be argued for the FCC in court by the Office of the General Counsel, would involve members of the General Counsel's office.

Under the Sunshine Act—which opened up many federal meetings to the public—most Commission meetings are open. Attending such a meeting can be of special value to the student studying communication management, policy, or law.

Outreach to the public by the Commission and its staff, in addition to the work of the FCC's Office of Public Affairs, is accomplished principally through attendance and participation as speakers and on panels at conventions of communication, industry, and citizen associations. On occasion the Commission holds regional meetings on important items under consideration for rule-making.

The following chart lists all the commissioners who have been on the FCC from 1934 to the present, with their dates of service:

Federal Communications Commission

Commissioner	Party	State affiliation	Term of service
*Eugene O. Sykes	D	Misissippi	July 11, 1934 - Apr. 5, 1939
Chairman			July 11, 1934 - Mar. 8, 1935
*Thad H. Brown	R	Ohio	July 11, 1934 - June 30, 1940
*Paul A. Walker	D	Oklahoma	July 11, 1934 - June 30, 1953
Acting Chairman			Nov. 3, 1947 - Dec. 28, 1947
Chairman			Feb. 28, 1952 - Apr. 17, 1953
*Norman Case	R	Rhode Island	July 11, 1934 -June 30, 1945
Irvin Stewart	D	Texas	July 11, 1934 - June 30, 1937

Commissioner	Party	State affiliation	Term of service
*George Henry Payne	R	New York	July 11, 1934 - June 30, 1943
*Hampson Gary	D	Texas	July 11, 1934 - Jan 1, 1935
*Anning S. Prail	D	New York	Jan. 17, 1935 - July 23, 1937
Chairman			Mar. 9, 1935 - July 23, 1937
*T.A.M. Craven	D	Dist of Col. Virginia	Aug. 25, 1937 - June 30, 1944
*Frank McNinch	D	N. Carolina	Oct. 1, 1937 - Aug. 31, 1939
Chairman			Oct. 1, 1937 - Aug. 31, 1939
*Frederick Thompson	D	Alabama	Apr. 13, 1939 - June 30, 1941
*James Lawrence Fly	D	Texas	Sept. 1, 1939 - Nov. 13, 1944
Chairman			Sept. 1, 1939 - Nov. 13, 1944
*Ray C. Wakefield	R	California	Mar. 22, 1941 - June 30, 1947
*Clifford J. Durr	D	Alabama	Nov. 1, 1941 - June 30, 1948
*Ewell K. Jett	Ind	Maryland	Feb. 15, 1944 - Dec. 31, 1947
Intern Chairman			Nov. 16, 1944 - Dec. 20, 1944
*Paul A. Porter	D	Kentucky	Dec. 21, 1944 - Feb. 25, 1946
Chairman			Dec. 21, 1944 - Feb. 25, 1946
Charles R. Denny	D	Dist of Col.	Mar. 30, 1945 - Oct. 31, 1947
Acting Chairman			Feb. 26, 1946 - Dec. 3, 1946
*William H. Willis	R	Vermont	July 23, 1945 - Mar. 6, 1946
Rosel H. Hyde	R	Idaho	Apr. 17, 1946 - Oct. 31, 1969
Chairman			Apr. 18, 1953 - Apr. 18, 1954
Acting Chairman			Apr. 19, 1954 - Oct. 3, 1954
Acting Chairman			May 1, 1966 - June 26, 1966
Chairman			June 27, 1966 - Oct. 31, 1969
*Edward M. Webster	Ind	Dist of Col.	Apr. 10, 1947 - June 30, 1956
*Robert F. Jones	R	Ohio	Sept. 5, 1947 - Sept. 19, 1952
*Wayne Coy	D	Indiana	Dec. 29, 1947 - Feb. 21, 1952
Chairman			Dec. 29, 1947 - Feb. 21, 1952
George E. Sterling	R	Maine	Jan. 2, 1948 - Sept. 30, 1954
*Frieda B. Hennock	D	New York	July 6, 1948 - June 30, 1955
*Robert T. Bartley	D	Texas	Mar. 6, 1952 - June 30, 1972
*Eugene H. Merrill	D	Utah	Oct. 6, 1952 - Apr. 15, 1953
John C. Doerfer	R	Wisconsin	Apr. 15, 1953 - Mar. 10, 1960
Chairman			July 1, 1957 - Mar. 10, 1960
Robert E. Lee	R	Illinois	Oct. 6, 1953 - June 30, 1981
Interim Chairman			Feb. 5, 1981 - Apr. 12, 1981
Chairman			Apr. 13, 1981 - May 18, 1981
*George McConnaughey	R	Ohio	Oct. 4, 1954 - June 30, 1957
Chairman			Oct. 4, 1954 - June 30, 1957
*Richard A. Mack	D	Florida	July 7, 1955 - Mar. 3, 1958
*Frederick W. Ford	R	W. Virginia	Aug. 29, 1957 - Dec. 31, 1964
Chairman			Mar. 15, 1960 - Mar. 1, 1961

Commissioner	Party	State affiliation	Term of service
*John S. Cross	D	Arkansas	Mar. 23, 1958 - Sept. 30, 1962
Charles H. King	R	Michigan	July 19, 1960 - Mar. 2, 1961
Newton N. Minow	D	Illinois	Mar. 2, 1961 - June 1, 1963
Chairman			Mar. 2, 1961 - June 1, 1963
E. William Henry	D	Tennessee	Oct. 2, 1962 - May 1, 1966
Chairman			June 2, 1963 - May 1, 1966
Kenneth A. Cox	D	Washington	Mar. 26, 1963 - Aug. 31, 1970
Lee Loevinger	D	Minnesota	June 11, 1963 - June 30, 1968
James J. Wadsworth	R	New York	May 5, 1965 - Oct. 31, 1969
Nicholas Johnson	D	Iowa	July 1, 1966 - Dec. 5, 1973
H. Rex Lee	D	Dist of Col.	Oct. 28, 1968 - Dec. 31, 1973
Dean Burch	R	Arizona	Oct. 31, 1969 - Mar. 8, 1974
Chairman			Oct. 31, 1969 - Mar. 8, 1974
Robert Wells	R	Kansas	Nov. 6, 1969 - Nov. 1, 1971
Thomas J. Houser	R	Illinois	Jan 6, 1971 - Oct. 5, 1971
Charlotte T. Reid	R	Illinois	Oct. 8, 1971 - July 1, 1976
Richard E. Wiley	R	Illinois	Jan. 5, 1972 - Oct. 13, 1977
Chairman			Mar. 8, 1974 - Oct. 13, 1977
Benjamin L. Hooks	D	Tennessee	July 5, 1972 - July 25, 1977
James H. Quello	D	Michigan	April 30, 1974 -
Glen O. Robinson	D	Minnesota	July 10, 1974 - Aug. 30, 1976
Abbott M. Washburn	R	Minnesota	July 10, 1974 - Oct. 1, 1982
Joseph R. Fogarty	D	Rhode Island	Sept. 17, 1976 - June 30, 1983
Margita E. White	R	Sweden	Sept. 23, 1976 - Feb. 28, 1979
Charles D. Ferris	D	Mass.	Oct. 17, 1977 - Apr. 10, 1981
Chairman			Oct. 17, 1977 - Feb. 4, 1981
Tyrone Brown	D	Virginia	Nov. 15, 1977 - Jan. 31, 1981
Anne P. Jones	R	Mass.	Apr. 7, 1979 - May 31, 1983
Mark S. Fowler	R	Canada	May 18, 1981 - Apr. 17, 1987
Chairman			May 18, 1981 - Apr. 17, 1987
Mini Weyforth Dawson	R	Missouri	July 6, 1981 - Dec. 3, 1987
Henry M. Rivera	D	New Mexico	Aug. 10, 1981 - Sept. 15, 1985
Stephen A. Sharp	R	Ohio	Oct. 4, 1982 - June 30, 1983
Dennis R. Patrick	R	California	Dec. 2, 1983 - Aug. 7, 1989
Chairman			Apr. 18, 1987 - Aug. 7, 1989
Patricia Diaz Dennis	D	New Mexico	June 25, 1986 - Sept. 29, 1989
Alfred C. Sikes	R	Missouri	Aug. 8, 1989 -
Chairman			Aug. 8, 1989 -
Sherrie P. Marshall	R	Florida	Aug. 21, 1989 -
Andrew C. Barrett	R	Illinois	Sept. 8, 1989 -
Ervin S. Duggan	D	Georgia	Feb. 28, 1990 -

*Deceased

STAFF OFFICES

The staff Offices and operating Bureaus of the FCC are organized into functional segments, principally into Divisions, Branches, and Sections. Each of these offices has a Chief, who supervises the work of his or her staff and reports, in turn, to the Branch Chief, Division Chief, or Bureau Chief. In some instances the special functions of a given office may result in that office, in practice, bypassing the nominal organizational route and reporting directly to a higher level, or to a commissioner who has special Commission-wide responsibility for the substantive area of that office. The organizational breakdowns, offices, and responsibilities discussed in this book are based on the FCC's structure at the beginning of the 1990s. Under the direction of the Managing Director, and with the approval of the Commission, the composition and duties of any office or bureau or its subdivisions may change at any time.

Office of Managing Director

The Managing Director is appointed by the Chairman with the approval of the Commission and serves as the administrative head of the FCC, assisting the chairman in carrying out the latter's responsibilities. The Managing Director formulates all management policies and supervises the Commission's bureaus and offices with respect to management and administrative matters, but not in relation to any substantive regulatory matters. The Managing Director deals with personnel, labor-management relations, budget and finance, information processing, procurement, office space, office services, supplies and property, records, security, and emergency communications.

In addition, the Managing Director provides audio visual services for the Commission, deals with ex parte matters (communications received from the public outside of the designated process for a given matter under consideration), coordinates international telecommunications activities, including travel fund allocations, and supervises the FCC library. Further, the Managing Director serves as the Commission's Defense Director, works with the General Counsel on license fees, and administers a number of special short-term and long-term management efficiency programs (see Figure 2).

The Office of Managing Director has seven major suboffices to deal with its broad responsibilities.

Management Planning and Program Evaluation Office   This office develops and implements the FCC internal program evaluation systems; manages organizational planning, operations, procedures, and work-flow analysis; and supervises management and productivity improvement. It prepares key FCC reports, such as the Annual Report to Congress on Goals, Objectives, Priorities, and Accomplishments, and the Managing Director's annual Report to the Chairman on Accomplishments. It coordinates the management of the various outside advisory committees that the FCC appoints from time to time. It is the administrative office for Emergency Broadcast System (EBS) functions under the FCC.

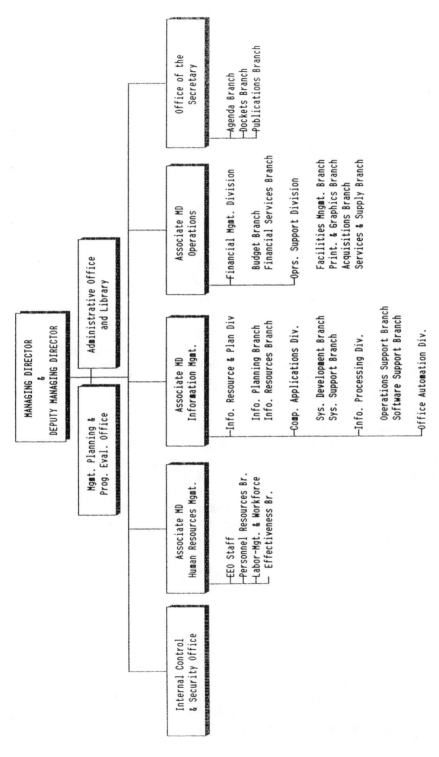

▲ *Figure 2   Office of Managing Director organizational chart.*

Library   In addition to acquiring and maintaining an extensive collection of books, periodicals, journals, court citation files, legislative histories, and other materials relating to communications, the FCC Library provides special assistance to the Commission staff in researching any given topic. It offers assistance to members of the public, as well, who wish to use the Library's materials. It also has continuing liaison and a materials exchange program with other government and nongovernment libraries.

Internal Control and Security Office   This office is responsible for the security of personnel and documents, as well as for the physical plant. It handles employee conflict of interest matters, including the Commission's Financial Disclosure requirements for certain personnel. It is in charge of employee clearance for access to classified information, attendance at international meetings, and visits to restricted sites. It administers the FCC's ex parte rules, and handles Freedom of Information requests to the FCC.

Associate Managing Director for Human Resources Management   This office consists of the *Equal Employment Opportunity (EEO) Staff*, *Personnel Resources Branch*, and the *Labor-Management and Workforce Effectiveness Branch*. It is responsible for all FCC programs related to the management of human resources, and for the internal EEO program. It handles occupational health and safety, and drug-testing programs. Recruitment, classification, merit pay, performance evaluation, and other employee development and training programs are included. Its recruitment includes liaison with colleges and universities to seek eligible candidates for FCC positions. It administers the employees' benefits programs, such as health and life insurance.

Associate Managing Director for Information Management   This office contains an *Information Resource and Planning Division* with its *Information Planning* and *Information Resources Branches*; *Computer Applications Division*, with *Systems Development* and *Systems Support Branches*; *Information Processing Division*, with *Operations Support* and *Software Support Branches*; and an *Office Automation Division*. Its principal duties are to manage all records and forms in the most efficient manner possible and to work on ways of reducing paperwork for licensees and for the Commission.

Associate Managing Director for Operations   This office has a *Financial Management Division* with a *Budget and Financial Services Branch*, and an *Operations Support Division* with *Facilities Management*, *Print and Graphics*, *Acquisitions*, and *Service and Supply Branches*. Its principal responsibilities relate to budgeting, financial management, relations with congressional appropriations committees, and efficiently managing the FCC's facilities and other support services.

Office of the Secretary   The Secretary is the keeper of the FCC's official seal and of official records of actions taken by the Commission. The Secretary signs FCC orders and other documents, and certifies those that are to be published in the Federal Register or released to the public. This office has *Agenda, Dockets, and Publications*

Branches. The *Agenda Branch* processes all agenda materials for Commission meetings, and maintains the records of all official actions taken. The *Dockets Branch* collects and prepares official records, called public dockets, and keeps history card files of all hearing and rule cases for reference use by the Commission staff and by the public. It maintains the minutes of Commission meetings, for internal and public use. The *Publications Branch* prepares documents for publication in official sources, including the *Federal Register*.

Office of the General Council

The General Counsel advises the FCC on legal issues involved in establishing and implementing policy, handles legal questions affecting the agency's internal operations, coordinates the preparation of the FCC's legislative program, and represents the Commission in court. This is accomplished through *Litigation, Adjudication,* and *Administrative Law Divisions* (see Figure 3).

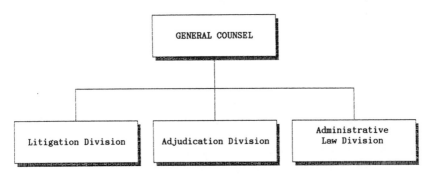

▶ *Figure 3   General Council organizational chart.*

Litigation Division   This office advises the FCC on all legal questions pertaining to proposed actions and policies in light of past and present court decisions and cases. It represents the Commission before the U.S. district courts and also before the Supreme Court in any proceedings in which the FCC is directly involved or in which it has an interest.

Adjudication Division   This division analyzes all FCC proposed Memorandum Opinions and Orders (that is, final action on items before the Commission) and the testimony and documents pertaining to hearings to determine if the decisions are consistent with the FCC's Rules and Regulations. (Hearings are held when a party appeals a Commission action.) It reviews decisions of the Review Board and, in some instances, of the initial adjudicatory body, the Administrative Law Judges.

Administrative Law Division   This division is involved with a number of Commission-wide matters, including participation in international conferences and the implementation of international communication agreements, works with the Common Carrier Bureau and the Office of Engineering and Technology on satellite matters, deals with the FCC's legal national defense responsibilities, administers the legal aspects of Experimental Radio and Industrial, Medical and Scientific services frequency devices, participates in frequency allocation and other rule-making affecting more than one bureau, and advises the Managing Director on the Freedom of Information, Privacy, and Sunshine Acts.

Office of Administrative Law Judges

This office conducts formal hearings relating to investigations, rule-making, and adjudications. It acts on motions, petitions and other pleadings filed in formal FCC proceedings. It functions as judges do in nonjury cases. Its decisions are subject to review by the FCC's Review Board.

Review Board

The Board hears appeals from decisions issued by Administrative Law Judges (ALJ) and issues a final ruling, subject to further review by the Commission. In some instances the Commission decides that it will itself hear the appeal from the ALJ decision, bypassing the Review Board.

Office of Legislative Affairs

This office implements the FCC's legislative programs by informing Congress of the Commission's regulatory decisions, responding to congressional inquiries, and recommending or commenting on proposed changes in existing law which affect the FCC or its processes. It works with the Managing Director in the preparation of the FCC's annual report and annual budget proposals to Congress, coordinates appearances of the Chairman and Commissioners before congressional committees, and serves as liaison between members of Congress on behalf of their constituents and various Commission offices.

Office of Plans and Policy

The Office of Plans and Policy (OPP) is the principal adviser to the Commission on both short-term and long-term economic and technical policies, including sociological and educational implications. OPP coordinates all of the FCC's policy research and development, and recommends budgets and priorities for these programs. It manages the accounts for all outside contract research studies. OPP coordinates the FCC's domestic communications positions, represents the agency at appropriate interagency meetings, and participates in the development of international communications policy. It prepares briefings, position papers, agenda items, and proposed Commission actions that deal with policy. The Chief of OPP reports directly to the Chairman.

Office of Engineering and Technology

The Chief Engineer heads the Office of Engineering and Technology (OET) and is the FCC's adviser on technical, engineering, and scientific matters, both in domestic regulation and international matters. OET's broad responsibilities include the development of information on all areas of communication techniques and equip-

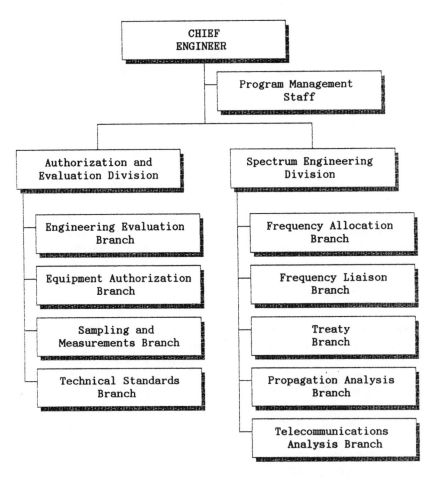

▶ *Figure 4*   *Office of Engineering and Technology organizational chart.*

ment; studying terrestrial and space communications; advising all offices in the FCC on spectrum management, frequency allocation and technical standards; maintaining liaison with other agencies, foreign governments, and the public on technical matters; preparing appropriate legislative recommendations; and participating in the formulation of regulations, agenda items, and other proceedings in regard to engineering and technical matters, including type-accepted equipment. OET performs these function through three major offices: the *Program Management Staff*; the *Authorizations and Evaluation Division* and its *Engineering Evaluation, Equipment Authorization, Sampling and Measurements*, and *Technical Standards Branches*; and the *Spectrum Engineering Division*, which contains *Frequency Allocation, Frequency Liaison, Treaty, Propagation Analysis*, and *Telecommunications Analysis Branches* (see Figure 4).

Program Management Staff   This office directs OET's internal administrative and management functions, coordinates surveys and analyses, estimates the budgets, and is liaison on these matters with Congress and with the General Accounting Office (GAO).

Authorization and Evaluation Division   This division formulates rules, technical standards and general policy for communications equipment and devices. It represents the FCC on interagency, national, and international committees dealing with technical standards. It's *Equipment Authorization Branch* tests and evaluates all types of equipment, makes recommendations regarding type acceptance, processes applications for equipment approval, and monitors equipment operation at selected sites. The *Sampling and Measurements Branch* develops and performs tests and provides data on equipment for new rules and regulations; designs, constructs, evaluates, and applies new electronic test equipment; and provides electronic measurement standards for the FCC, other agencies, and the industry as a whole. It monitors communications equipment to ensure compliance with the FCC Rules. The *Technical Standards Branch* conducts studies to develop technical rules and standards for various services, and advises the FCC and Congress on equipment interference and compatibility. The *Engineering Evaluation Branch* conducts laboratory and field studies to evaluate the spectrum efficiency of new or modified equipment.

Spectrum Engineering Division   This division coordinates with the National Telecommunications and Information Administration (NTIA) on frequency allocation matters, maintains liaison on spectrum use and interference control with international organizations, including notification to the International Telecommunications Union (ITU) on U.S. frequency assignments. It studies and plans future use of the spectrum, coordinates FCC rule-making on any changes in the U.S. Table of Frequency Allocations, and studies ways of improving the efficiency of spectrum use. The *Frequency Allocations Branch* studies spectrum utilization and recommends regulatory actions regarding its long-term uses, including frequency allocations, studies new technical developments in terms of their impact on spectrum utilization, and maintains liaison with the public on spectrum management problems. The *Frequency Liaison Branch* maintains FCC contact with the International Radio Advisory Committee (IRAC) and with other government agencies on spectrum use. It is also in charge of the Experimental Radio Service, developing standards and forms, processing applications, issuing licenses, monitoring and enforcing compliance with the rules, and gathering data on experiments for rule-making purposes. The *Treaty Branch* participates in international meetings, develops proposals for international communications treaties, administers U.S. requirements regarding frequency assignments under those treaties, is a clearinghouse for all information on international frequency use, and assigns call letters to government agencies. The *Propagation Analysis Branch* studies the effects of the ionosphere, climate, meteorological occurrences, and topography on radiowave propagation; studies equipment, modulation, propagation, environmental, and operational problems of terrestrial telecommunications; designs experiments to solve propagation problems; and studies and evaluates potential hazards and health risks from radiofrequency radiation. The *Telecommuni-*

cations Analysis Branch studies new telecommunications technologies' potential impact on the industry and on national and international communications systems, and makes recommendations for appropriate FCC regulations. It also studies spectrum efficiency and the performance of digital and analog telecommunications systems, and recommends improvements for such systems. In addition, it assists other OET offices in the development of mathematical and computer models for evaluating spectrum efficiency and interference issues. It functions as an education office within OET on developing the staff's technical proficiency and on the acquisition and use of automatic data-processing equipment for scientific computing.

Office of Public Affairs

The Office of Public Affairs (OPA), is responsible for the FCC's news media, consumer assistance and small business, and minority enterprise programs. OPA informs the public of the FCC's regulatory requirements, of the process for public participation, and of the FCC's policies regarding minority participation in the telecommunications field. It publishes daily news releases, public notices, and other informational material, including bulletins on different organizational aspects and procedures of the FCC, handles requests for information, and provides internal informational services. It prepares the FCC's Annual Report. Specific publications and other informational material available to the public are detailed in Chapter 6. OPA is the FCC's principal contact with minority organizations, industry representatives, and local, state, and other federal agencies in relation to minority affairs, and makes recommendations to the Commission on minority enterprise rule-making. The Office of Public Affairs has a *News Media Division* and a *Consumer Assistance and Small Business Division*, plus an *Assistant Director of Minority Enterprise* (see Figure 5).

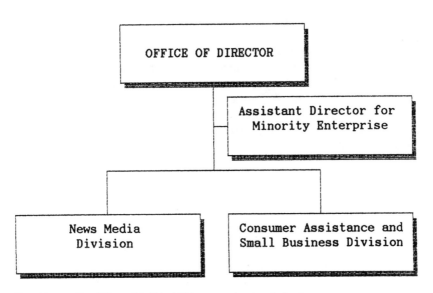

▶ *Figure 5   Office of Public Affairs organizational chart.*

Assistant Director of Minority Enterprise   The extent of the FCC's efforts in this area depends on the philosophy of the party in power at any given time. During the Reagan deregulatory period of the 1980s, for example, FCC activities in promoting minority participation in telecommunications were greatly modified. The Assistant Director's principal responsibilities are to establish contacts with minority parties and provide advice and guidance on entry into, and participation in, the telecommunications industry.

News Media Division   This division disseminates news and official announcements of FCC activities, including opinions and orders, public notices, fact sheets, news bulletins, and other documents; maintains liaison with print and electronic media; coordinates press conferences and interviews and media appearances of FCC staff members; and provides a daily clipping service to FCC offices on current telecommunications developments. It also does research for the preparation of materials for speeches, conferences, and meetings, prepares the Annual Report to Congress, and writes and edits articles for publication on FCC activities and programs. It has a separate physical office for press release and public notice distribution.

Consumer Assistance and Small Business Division   This division maintains liaison with consumer organizations, informing the Commission on public concerns and informing the public on FCC policies and procedures and public participation. It works with other FCC offices in developing bulletins and publications for the public, and has an information center to provide assistance in person, by phone, or through the mail. It assists the Assistant Director for Minority Enterprise in disseminating information.

Office of the Inspector General

The Inspector General conducts audits and investigations to discover and remedy waste, fraud, and other abuses, and looks into reported instances of misconduct of FCC employees—Commissioners and staff alike. It reviews government-wide proposed legislation and regulations relating to the FCC and recommends policies to prevent fraud or other abuses and to promote the most efficient operation of the FCC. It reports any serious problems immediately to the Chairman and, within seven days, to Congress, and recommends corrective actions. It files semiannual reports to the Chairman and to Congress on any problems or deficiencies in the administration of the FCC.

3

▼
▼
▼
▼
▼

Organization and Operation: Operating Bureaus

THE MASS MEDIA BUREAU

For many years known as the Broadcast Bureau, in the 1980s the name was changed to reflect the growing number of nonbroadcast mass media services that also came under the jurisdiction of the Bureau. While other bureaus may deal with many more licensees than does the Mass Media Bureau — for example, the Private Radio Bureau licenses considerably more business radio operations than all of the commercial and noncommercial radio and television stations combined — most Americans have a more direct relationship to broadcast stations than to any other service and think only of television and radio when the FCC is mentioned.

The Mass Media Bureau regulates AM, FM, and television broadcast stations, Direct Broadcast Satellite (DBS), Instructional Television Fixed Service (ITFS), and related facilities, and administers and enforces cable TV rules, including licensing cable relay private microwave radio facilities. It processes applications and renewals for licenses, develops proposed rules and regulations relating to the electronic media, and administers U.S. obligations under international treaties pertaining to these media. It consists of an Administration and Management Staff, an Authorization Programming Group, and four Divisions:

1. Audio Services, which includes three branches dealing respectively with AM, FM, and auxiliary radio services. This division also has a data management staff and a public reference room.
2. Video Services, with four branches dealing respectively with television stations, low-power television, cable, and distribution services. It has, in addition, a staff responsible for station ownership issues.
3. Policy and Rules, with four branches, legal, engineering policy, policy analysis, and allocations. In addition, it has an international matters staff. It works on proposed rules and regulations agenda items for the Commission relating to mass media matters, including frequency allocations.
4. Enforcement, with four branches handling complaints and investigations, equal employment opportunity, fairness/political programming, and hearing. (See Figure 6.)

Administration and Management Staff This staff is responsible for the administrative and management functions within the bureau, including personnel, equipment, facilities, training, records, and budget.

143,168

LIBRARY

25 ▲

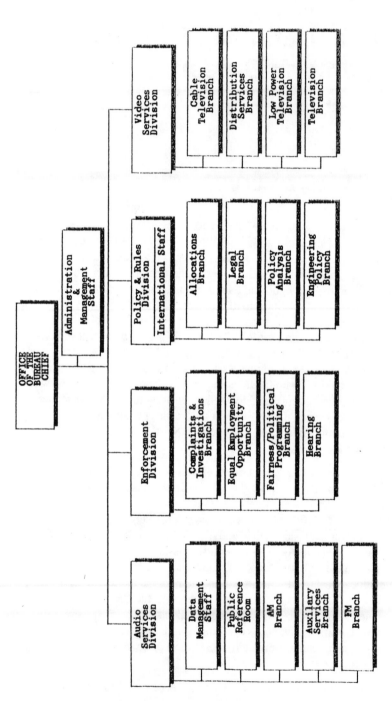

▶ *Figure 6   Mass Media Bureau organizational chart.*

Authorization Programming Group   This office handles systems analysis matters for the bureau, including data base design, implementation, and control.

Audio Services Division   This division processes all applications for commercial and noncommercial AM and FM radio facilities, including new, modifications, renewals, and transfers.

Its *AM Branch* and *FM Branch* functions are identical for the different services. They review all applications in terms of legal, technical, and financial qualifications, and service to the public; interpret the rules pertaining to these services; and issue temporary authorizations for equipment testing, power changes, experimental testing, and suspension or emergency resumption of operations.

The *Auxiliary Services Branch* has the same responsibilities for FM translator stations and associated facilities such as remote pickups, studio-transmitter links (STL), and intercity relays.

The *Data Management Staff* maintains computerized databases for all the services processed, and its Public Reference Room has all files, records, and correspondence regarding these services available for public inspection.

Video Services Division   This division processes all applications for commercial and noncommercial television facilities, including new, modifications, renewals, and transfers.

Its *Television Branch* reviews all applications in terms of legal, technical, and financial qualifications, and service to the public; interprets the rules pertaining to television services; and issues temporary authorizations for equipment testing, power changes, experimental testing, and suspension or emergency resumption of operations.

The *Low-Power Television Branch* processes all applications for LPTV and TV translator stations; interprets the rules for these services; issues temporary authorizations; and maintains computerized databases for all bureau television services. Because mutually exclusive (MX) applications (more than one applicant for the same channel) are settled by lottery, the branch decides which applications will go into the lottery and selects tentative winners. In cases of appeal, the branch prepares the appropriate legal, engineering, accounting, and economic documents for a hearing before the Administrative Law Judges, and presents oral arguments before the ALJ and the Commission.

The *Distribution Services Branch* is responsible for processing applications for DBS and ITFS, interprets the rules, and issues temporary authorizations, as do the previously noted branches for their services. This branch also coordinates DBS applications with the Private Radio Bureau in order to monitor possible interference among microwave services.

The *Cable Television Branch* maintains a file of cable television registrations and annual reports, and processes applications for cable relay services and cable use of aeronautical frequencies.

The *Ownership Section* of the division examines station ownership reports that are required to be filed with new applications, annually from licensees, with renewals and transfers of control, and with ownership changes.

Policy and Rules Division   This office studies and prepares rule-making relating to the electronic mass media, including their legal, social, technical, and economic aspects. It assists the Commission in preparing for international conferences, conducts rule-making proceedings relating to the FM and TV tables of channel assignments, and prepares reports on the various legal, technical, and economic aspects of the mass media industry.

Its *Legal Branch* processes petitions for rule-making, prepares rule-making proposals and Notices of Inquiry (NOI—see Chapter 5 of this book), recommends proposed legislation, prepares special reports to Congress, interprets the legal aspects of the Rules and Regulations, studies appropriate aspects of network operations, and provides legal research and information on the electronic media to other Commission offices.

The *Engineering Policy Branch* processes petitions for rule-making regarding technical aspects of the electronic media, conducts studies relating to rule-making, recommends regulations and legislation, interprets the engineering aspects of the rules, supplies engineering information for reports to Congress, prepares recommendations on frequency requirements for the electronic media, and negotiates conflicts between U.S. and foreign broadcast station assignments.

The *Policy Analysis Branch* processes petitions relating to the economic aspects of the electronic media; provides economic data, including cost-benefit analyses, for proposed rule-making; compiles economic, social, and programming data from licensees; recommends legislation; conducts studies relating to the long-range needs of educational broadcasting and serves as liaison with educational/public broadcasting stations and organizations.

The *Allocations Branch* processes petitions and initiates and conducts rule-making proceedings related to the amending of the FM and TV Tables of Assignments.

The *International Staff* processes petitions, conducts studies, and analyzes and initiates rule-making related to international aspects of the electronic media rules and regulations. It provides engineering and legal support for negotiation, interpretation, and enforcement of international agreements, and provides and receives notification to and from foreign governments on station assignments.

Enforcement Division   This division handles formal and informal complaints against stations and cable systems, conducts investigations as appropriate, administers the equal-time requirements for political candidates, and deals with station and cable employment data. The division also handles Freedom of Information requests pertaining to the Bureau.

The *Complaints and Investigations Branch* evaluates complaints from all sources, including the public, industry, and Congress, and disposes of them by correspondence, investigation, and dismissal or sanctions. It maintains a complaint file for public inspection—a folder for each station and cable system, containing all comments, correspondence, and records of Commission actions.

The *Fairness/Political Programming Branch* processes complaints, comments, and inquiries regarding Sections 315 and 312(a)(7) of the Communications Act of 1934, which deal with equal time on broadcast stations for political candidates. The office makes interpretations and rulings regarding station compliance. This subject is dealt with in more detail later in this chapter. While the Fairness Doctrine has been abolished, certain aspects of the doctrine developed in court cases still apply, and this branch is responsible for interpreting and disposing of complaints and inquiries on this matter.

The *Hearing Branch* is responsible for hearing matters involving AM, FM, TV, translators, distribution systems, cable systems, and other Mass Media Bureau services. It conducts pretrial investigations, prepares all pleadings, and presents evidence, examines witnesses, and makes oral arguments before the Administrative Law Judges, the Review Board, and the Commission.

The *Equal Employment Opportunity Branch* reviews all applications and petitions in regard to EEO regulations, receives and studies employment reports and affirmative action plans, investigates charges of discrimination, maintains liaison with the federal EEO Commission, assists stations in regard to EEO regulations, and recommends changes in FCC EEO Rules and Regulations.

Key Mass Media Regulatory Areas

A number of FCC mass media procedures, policies, and rules apply to all stations, both radio and TV.

Frequency Assignments and Licensing   Within the frequency bands allocated for radio and TV broadcasting, the FCC assigns specific stations specific frequencies and power. The chief consideration is to avoid interference with other stations—the condition that required the establishment of the first broadcast regulatory body, the Federal Radio Commission, in 1927. When an applicant applies for a new station or for a change in its current station's operations, it first must receive a construction permit (CP), and when the construction is completed and the station shows it is operating as it proposed, it can be granted a license. During the period of license—7 years for radio stations, 5 years for television stations—the FCC expects the station to operate in accordance with the FCC rules and the provisions of its license authorization. The FCC may modify a license, assign call letters, license transmitter operators, process requests for a transfer of license, and decide on renewal of license applications. At one time the latter was a key factor in motivating station performance. A station was obligated to submit detailed information on how it had served the public interest through programming and other services, and public interest groups were allowed to participate in the renewal process. Under the Reagan deregulatory policies of the 1980s, renewal applications were cut down to postcard size, and stations are no longer required to prove their public interest service; absent obvious or gross violations of the rules, stations are almost automatically given license renewals. Radio licensing is under the jurisdiction of the Audio Services Division, television licensing under the Video Services Division. New or changed frequency allocations are handled by the Engineering Policy and Allocations Branches of the Policy and Rules Division.

Censorship   While many broadcasters believe that many of the FCC rules and policies constitute censorship and a violation of First Amendment rights, and much of the public believes that the FCC has the right, if not the obligation, to decide on program content, the Communications Act of 1934, as amended, specifically forbids the FCC from censoring programs. Section 326 of the Act states:

> Nothing in this Act shall be understood or construed to give the Commission the power of censorship over the radio communications or signals transmitted by any radio station, and no regulation or condition shall be promulgated or fixed by the Commission which shall interfere with the right of free speech by means of radio communication.

Indecency   Complaints about program content and, conversely, charges of censorship come to the Complaints and Investigations Branch of the Enforcement Division. Many of these deal with what the viewer or listener considers obscene or profane presentation. The FCC defines broadcast indecency as language or material that "depicts or describes, in terms patently offensive as measured by contemporary community standards for the broadcast medium, sexual or excretory activities or organs." The FCC has taken action against stations carrying such materials when it has believed that there was reasonable risk that children might have been among the audience. While neither the FCC nor the courts or any other governmental body has defined "contemporary community standards" for all of the United States, as this is being written a Congress-mandated 24-hour ban on so-called indecent broadcasting is in effect, although it is being challenged in the courts.

Political Broadcasting   Many people confuse the political "equal time" rule with the Fairness Doctrine. The latter, although no longer in existence as this is written in 1990, will be discussed briefly below. The political equal-time provision does exist, under Section 315 of the Communications Act. It states, in essence, that any station that permits a legally qualified candidate for a given office to use its broadcast facilities, on a paid or unpaid basis, must provide "equal opportunities" for use of the same facilities to all other legally qualified candidates for that office. News and public affairs programs are exempt. Stations may not censor any political broadcasts.

Each station must have a policy concerning sale of time to candidates for political advertisements; that policy must provide equal opportunity for all certified candidates to purchase time at least equal to the time purchased by any one candidate. Stations, however, do not have to offer free time to a candidate whose opponent may have purchased a great deal of advertising time, but who does not have enough campaign monies to match the opponent's paid advertising. It is this factor that has prompted many critics to state that money buys elections. That is, whoever has the most money can purchase enough broadcast time to get name recognition which, in turn, provides high rankings in polls, and, also in turn, leads to more attention from the news media. This leads to even more campaign donations for more ads, establishing that candidate as one of the front-runners, resulting in more free coverage than that candidate's opponents in news and public affairs programs. With rare exceptions, such a candidate is victorious in the primary and the election. The candidate with less money, therefore may be ruled out of contention by the media, regardless of qualifications.

Section 315 also requires stations to sell political time at the "lowest unit charges," and to keep a file of requests for political time, with the disposition of those requests. Another part of the Communications Act, Section 312(a)(7), provides candidates for federal elective office the right of reasonable access to broadcast media.

During the deregulatory period of the 1980s parts of the equal-time rule were modified so as to favor the front-running candidates and make it possible for stations to preclude from some coverage candidates they did not favor. News programs had been exempt from the equal time rule, and a broadening of the concept of news has made it possible for stations to carry so-called public affairs interviews and other features with some candidates and to omit from coverage other candidates for the same office. Further, where stations had previously been prohibited from initiating and conducting candidate debates—but were permitted to cover debates held by a non-station organization—stations now can hold their own debates, and in some instances they have chosen to invite some candidates for a given office, while excluding others for the same office (a situation that exists frequently where many persons are running in a primary election). The station therefore establishes the candidates of its choice as viable front-runners and virtually eliminates from the race candidates it does not like.

The Fairness Doctrine, which had dominated the controversy between many broadcasters and the FCC for many years, is not in existence at this writing. It was eliminated by the FCC in 1987 shortly after President Reagan vetoed a Fairness Doctrine bill that had been passed overwhelmingly by the House and the Senate, but which could not muster enough votes for veto override in the Senate. Essentially, application of the Fairness Doctrine meant that if a station presented only one side of an issue that was controversial in that station's community of service, and if the station refused to provide representatives of the other side opportunity (though not necessarily equal time) to reply, and if a complaint was made to the FCC, and if the FCC found the complaint to be valid, the FCC could require the station to provide coverage of the other side of the controversial issue. Some observers expect Congress to enact another Fairness Doctrine law in the early 1990s.

Matters relating to political programming or fairness in programming are handled by the Fairness/Political Programming Branch of the Enforcement Division.

Monopoly and Multiple Ownership   The *scarcity principle* in broadcasting recognizes the limit on the number of frequencies available for broadcast stations and the need, therefore, to promote and maintain diversity in station ownership and the dissemination of information and ideas to the public. To achieve that purpose the FCC is responsible for implementing several diversification regulations under the general anti-monopoly area: multiple ownership, duopoly, and cross-ownership.

No entity may own more than 12 AM, 12 FM, and 12 TV stations nationally, provided the television stations do not reach more than 25% of the country's television households. An exception to this rule is made for minority-owned stations, in which case a maximum of 14 in each service may be licensed to one owner, with a maximum of 30% of the television households reached. The rule was originally adopted in order to "maximize diversification of program and service viewpoints as well as prevent any undue concentration contrary to the public interest."A 7-7-7 limitation was changed to 12-12-12 in 1984, during the deregulation decade.

Duopoly   The duopoly rule enacted in 1970, prohibited one owner from having more than one TV, one AM, or one FM station in the same market. Existing AM-FM and TV-radio combinations were *grandfathered*—that is, allowed to continue—but the latter had to be divested when sold. In 1988 the FCC deregulated the rule to permit radio-UHF combinations in all markets and radio-VHF combinations in the top 50 markets, and relaxed the restrictions on common ownership of 2 radio stations in the same market.

Cross-ownership   The cross-ownership rule bars television or radio licensees from also owning and operating a daily English-language newspaper in the same community. Such combinations exist today because of grandfathering and recent FCC waivers of the rule.

Matters relating to monopoly are brought to the attention of the Complaints and Investigations Branch of the Enforcement Division, with the Legal Branch and Policy Analysis Branch of the Policy and Rules Division likely becoming involved if legal action or rule interpretations or changes become factors.

Station Management Responsibilities   Stations operations and programming are the responsibilities of the official licensees. A licensee may not be excused from adhering to the FCC *Rules and Regulations* by permitting another party to assume operational or programming decision-making at the station. The FCC does not monitor the day-to-day operations of the stations, such as charges for advertising time, employee relations (although the FCC does have equal opportunity guidelines), profits, personnel salaries, and similar internal affairs. Further, it licenses only the stations and the operators of station transmitters, and has no jurisdiction over other personnel such as announcers, disc jockeys, and other performers and artistic persons. The FCC does have some general rules regarding content, including obscenity and profanity as noted above, and if an on-air personality violates those rules, the FCC may hold the station directly responsible and levy a fine or even a suspension of license. In such instances the on-air personality who violated the rule is dealt with by the station, but not directly by the FCC. The Complaints and Investigations and the Equal Employment Opportunity Branches of the Enforcement Division frequently become involved in these areas.

Advertising   Advertising rates are the prerogative of the station, as is the option of whether to accept or reject any given item of programming, including advertising. All rates are negotiated by the individual station with its advertisers (usually through its station representative—that is, the organization that arranges for national and regional spots and sometimes, even for local ads). During the 1980s deregulation period previous limits on advertising time—program-length advertisements, designated at five minutes or more—were abolished, and there are now no limits on the amount of advertising a station may carry. A station must, however, identify any paid-for programming at the time it is broadcast, including the fact that it is sponsored, and by whom. The Complaints and Investigations Branch deals with violations in this area.

False or misleading advertising does not fall under FCC jurisdiction, but is the responsibility of the Federal Trade Commission (FTC). The FCC has no regulation concerning advertising that is offensive to some members of the public, except that if such advertising violates another FCC rule, such as that of profanity or indecency, the FCC may intervene. Federal law bans the advertising of cigarettes and little cigars on electronic media under FCC jurisdiction. Some people believe there is, as well, a ban on the advertising of hard alcoholic beverages; that is not so. The FCC may not censor any program matter, including advertisements. The absence of hard liquor ads on television and radio is not due to a government regulation but to a voluntary self-imposed ban by broadcasters.

Subliminal ads   Some viewers have complained to the FCC about subliminal advertising—words and pictures flashed briefly on the screen that are not consciously noted by the viewer, but are imprinted on the subconscious. The FCC bars such advertising, holding that it is deceptive and contrary to the public interest.

Payola, Rigged Quiz Shows, and Lotteries   Abuses by broadcasters over the years have resulted in federal law and FCC rules concerning these areas. Payola refers to illegal payments to disc jockeys to play and promote certain records. The FCC requires stations to announce when money or any other consideration is received for the presentation of any broadcast material. Quiz show scandals promoted legislation that makes illegal the presentation of programs purporting to be contests of knowledge or skill where the result is in any way prearranged. In 1975 Congress permitted stations to broadcast information on state-operated lotteries in the station's state or in adjacent states. The U.S. Criminal Code prohibits broadcasting of lotteries other than state-operated ones. In addition, while stations may sponsor contests or giveaways, the Communications Act forbids any predetermination of the outcome of that contest or deceiving the audience about the contest. An FCC rule requires that any station that holds a contest must fully disclose the terms of the contest and that the contest be conducted as announced. This area comes under the jurisdiction of the Complaints and Investigations Branch.

Call Letters   As the number of early radio licensees grew, stations east of the Mississippi River were given call signs beginning with *W*, while stations west of the Mississippi had theirs beginning with *K*. Stations in the east that already had calls beginning with *K* retained their call signs. Most early stations had only three-letter call signs; as the number of stations increased, four-letter identifications were assigned. Because many licensees operate more than one station with the same call sign in a community, many call letters are followed by -AM, or -FM, or -TV, to avoid confusion. The Audio Services Division and the Video Services Division handle call letter requests for their respective services.

Networks   The FCC does not license networks, only individual stations. However, the FCC exercises some degree of indirect regulation over networks because it has regulations pertaining to station affiliation agreements, programming policies, and other matters between a station and a network. By preventing stations from engaging in network agreements that it believes are harmful to the station or to the public interest, the FCC can affect the operations of the network. For many years the FCC had an Office of Network Study, to monitor potentially monopolistic and other negative practices by networks; however, this office was abolished in early 1980, in the pre-Reagan deregulation begun by the Carter administration. Violations of rules relating stations to networks are dealt with by the Complaints and Investigations Branch.

National Defense   In cooperation with the Federal Emergency Management Agency and other government agencies, the FCC established the Emergency Broadcast System (EBS), in which broadcast stations voluntarily participate. EBS facilities provide information to the public regarding any crisis, including war. It is used, as well, to inform the public of artificial disasters, and of natural disasters such as hurricanes that threaten life and property at the national, state, or local level.

Equal Employment Opportunity   The FCC rules state that equal opportunity in employment "shall be afforded by all licensees or permittees of standard, FM, television, or international broadcast stations . . . to all qualified persons, and no person shall be discriminated against in employment because of race, color, religion, national origin, or sex." Broadcast stations are required to file annual employment reports, with stations that have five or more full-time employees adding data on staff composition by job categories, and by gender and race. Applicants for new licenses or renewals who will have five or more employees must file with the FCC a model EEO program designed to assure equal employment opportunity for women and minority groups. This program is under the jurisdiction of the Equal Employment Opportunity Branch of the Enforcement Division.

The FCC and the U.S. Equal Employment Opportunity Commission (EEOC) have an agreement on procedures for receiving information, hearing complaints, and on other matters relating to EEO. The EEOC is responsible for administering laws barring discrimination because of age, and gender-based unequal pay for equal work. These laws have rarely been enforced since 1981.

AM Radio   AM radio is serviced by the AM Branch of the Audio Services Division. Domestic uses of frequencies are governed by international agreements on spectrum allocation. Within the 107 channels allocated for AM radio the United States has three major types: clear channels, regional channels, and local channels. Further, within these areas, there are four classifications based on power.

Clear channels   The FCC rules designate 47 clear channels. Each of these channels contains one or more class I stations that operate full time with 10 kw to 50 kw of power. Class I stations have interference priority; no other station may interfere with either their primary signal or their extended skywave signal, which is generated from dusk to dawn by bouncing off the Kenneally-Heaviside air layer and carrying hundreds and even a thousand and more miles. Class II stations may operate on the same channels with 250 watts to 50 kw of power, but may not interfere with the Class I stations. Some are able to operate full time, others part time, and many daytime only, going on the air at dawn and off at dusk so as not to cause interference to the Class I signal, which may be coming from a city in a distant part of the country.

Regional channels   Class III stations are assigned to 41 regional channels and operate with from 500 watts to 5 kw. Their principal purpose is to service a relatively small center of population and the surrounding rural areas. Some broadcast full time, others part time, and still others daytime only, in order to avoid interference with any other Class III stations on the same channel.

Local channels   Six remaining channels are assigned to Class IV stations, which operate with a maximum power of one kilowatt. Service areas of these stations are usually very small communities and the nearby countryside.

Daytime-only stations   More than half of the AM stations licensed in the United States are limited to daytime operation only. Clear channel stations are authorized to have both *groundwave* (or primary) service and *skywave* (or secondary) service. The reflection of the skywave signals off the ionosphere at night varies in distance and quality because of atmospheric conditions; nevertheless all these Class I stations are protected. Some daytime-only stations have special permission to go on the air before sunrise and operate up to two hours after sunset.

FM Stations   The FM Branch of the Audio Services Division is responsible for FM radio. Six classes of FM radio stations are assigned to three U.S. zones: zone I includes all or part of 18 northeastern states and the District of Columbia; zone 1-A covers southern California; zone II includes the rest of the country. Class A stations are low-power, with a maximum of 3 kw effective radiated power, and are assigned to all zones. Class B stations have a maximum of 50 kw, class B-1 a maximum 25 kw, and both are assigned to zones 1 and 1-A. Class C and C1 stations are 100 kw maximum, class C2 a maximum of 50 kw, and these are assigned to zone II. FM is a line-of-sight signal, and coverage areas are not as great as for AM, although the sound quality is considerably better. With maximum power and antenna height, class A stations reach about 15 miles, class B about 33 miles, and class C about 64 miles.

The FCC has a table allocating some 3000 commercial FM channels to about 2000 communities throughout the country. To avoid interference, the rules specify mileage separations between stations on the same or adjacent channels. An applicant for a new station must find a vacant channel in the table of allocations and prove to the FCC that putting it on the air with a specific power and antenna height will not cause interference to another station or result in interference to the new station.

Subsidiary service   Multiplexing—that is, using one or more of the FM signal's subchannels in addition to the main channel—permits an FM station to provide special subsidiary service in addition to its regular broadcasting. Operating at the same time as the regular broadcast signal, the subsidiary signal may carry foreign language programs, reading services for the blind, background music, specialized professional information, news reports, retail selling reports, and anything else for which there is a client. Stores, offices, manufacturing plants, physicians' offices, senior citizen centers, hospitals, and individuals at home with special converters contribute important income to the FM stations for the special audio services. In recent years the transmission of computer-type data such as stock market reports and financial news has grown on FM subsidiary channels.

Translators   FM translators receive off-the-air signals from the FM stations, and amplify and retransmit the signals on different channels that will not cause interference in the new geographical area. This enables FM stations to reach areas that would not otherwise receive the signal because of line-of-sight obstructions. Translators are miniature, unattended, remote-operated radio stations.

Television Stations   The Television Branch of the Video Services Division is responsible for television station services. Television channel allocations are similar to those of FM. Specific assignments are divided among several sets of frequencies: VHF channels 2–6, with a maximum power of 100 kw; VHF channels 7–13, with 316 kw maximum (higher on the frequency scale, these channels need more power to achieve coverage comparability with channels 2–6); and UHF channels 14–69, with 5000 kw maximum (even this much power does not provide UHF signal comparability with the VHF frequency bands). UHF occupies the frequencies from 470–806 MHz; VHF operates in the 54–88 MHz range for channels 2–6, and from 174–216 MHz for channels 7–13.

Although the FCC maintains that the UHF signal is now comparable to the VHF signal, practitioners will tell you that it isn't. VHF still has the more powerful signal. At the time UHF was introduced the FCC chose the path of intermixture, mixing VHF and UHF assignments in a community; some critics maintain that deintermixture, making any given community either all VHF or all UHF, would have resulted in greater comparability between the two television services.

For purposes of distance separation on the same channel to avoid interference, the country is divided into three geographic zones. In zone I, covering the Northeast and some of the Midwest, minimum co-channel separation is 170 miles between VHF stations and 155 miles between UHF. In zone II, covering virtually all of the rest of the country except for several states in the South and Southwest, the separation is 190 miles for VHF and 175 miles for UHF. Because of signal propagation characteristics near the Gulf of Mexico, zone II states Georgia, Alabama, Louisiana, Mississippi, and Texas are required to have a 220 mile separation for VHF and 205 miles for UHF.

Low power television   Television translators operate the same way FM translators do, as described above. TV translators enable stations to serve sparsely settled areas by retransmitting the original signal. In 1982 the FCC authorized the use of low power television stations—in effect, those that had been used as translators—as originating stations with power up to 10 watts for VHF and 1000 watts for UHF. This has resulted in more than 2000 low power TV stations either on the air or in the process of construction at the beginning of the 1990s, to serve limited, local needs of specified communities. The Low Power Television Branch in the Video Services Division handles this area.

Teletext/Videotex   The vertical blanking interval (VBI) of the video portion of the television signal is that portion that appears as a blank bar when the picture rolls. It is not visible to the eye of the viewer. Textual and graphic data may be put on the VBI and viewed with the aid of a special decoder; this is teletext, which opens up many special services to the television viewer such as closed captions for the hearing impaired, business information, special shopping reports, and other materials not usually received through one of the regular broadcast channels or even via a cable channel. When used in conjunction with a home personal computer, it is videotex, an interfacing that can provide many two-way services such as shopping, banking, library information, and health assistance.

Closed-captioning   The FCC has authorized special use of line 21 of the VBI for visual display of aural material accompanying TV programs. Hearing-impaired people can use this technical system; the VBI information does not interfere in any way with the viewing and listening of the larger audience.

Subscription TV   Subscription TV (STV) is synonymous with Pay-TV (PTV) or Pay-per-view (PPV). The FCC authorizes the transmission of scrambled signals that can be decoded only with a special device at the subscriber's set. During the 1980s deregulation period the FCC eliminated virtually all regulation of STV. While STV through conventional broadcasting has generally not been successful, it has grown considerably via cable.

Direct Broadcast Satellite   In 1982 the FCC authorized direct broadcast satellite (DBS) service, the broadcast of television signals directly to the home via satellite. A home satellite receiving dish is needed. The FCC assigns specific orbital slots and frequencies, based on a set of internally allocated frequencies, to each DBS licensee. While television stations have for many years sent their signals to affiliated stations via satellite, it wasn't until 1990 that plans were developed for a consortium of broadcast and cable organizations to send such programming directly to the home. Cable companies have already established satellite subscription services for individual homes. A key problem in satellite programming is piracy, where individuals buy or create decoders or unscramblers and pick up the signals without paying the required fees. Such piracy is in violation of FCC rules.

Educational/Public Broadcasting   Educational broadcasting is the term originally and still frequently used by the FCC for what most of the country calls Public Broadcasting. In the FCC *Rules and Regulations* it is also called noncommercial broadcasting. Educational or public or noncommercial stations may be applied for by nonprofit educational institutions or organizations, which must show that the programming will be, in general, of an educational nature. That could range from purely formal instructional programming to cultural materials to alternative popular programs (music, drama, etc.) not otherwise predominant on TV in a given community.

Noncommercial stations follow essentially the same licensing and renewal procedures and technical rules as do commercial stations. However, special rules and regulations do apply to noncommercial stations, in some instances having fewer bureaucratic requirements than for commercial stations, and in others more. For example, the multiple ownership rule does not apply to noncommercial educational stations, and some states have statewide public broadcasting systems with more television and radio stations than a commercial licensee would be permitted. Noncommercial stations, however, are not permitted to sell commercial time or carry paid advertisements. But during the 1980s deregulation period they were given greater leeway in announcing the names of products and, on TV, showing the logos of underwriters. Radio stations may carry non-paid commercial notices, with the prerogative of accepting general donations from commercial sources.

Beginning in 1962, when FCC Chairman Newton Minow established the Educational Broadcasting Branch, the FCC made special efforts to facilitate the development of public broadcasting stations. That ended in 1980, however, with the pre-Reagan deregulation by the Carter administration's FCC chairman, Charles Ferris, resulting in the abolition of the Educational Broadcasting Branch. Educational/public broadcasting matters are now a small part of the responsibilities of the Policy Analysis Branch. It should be noted that the FCC has no direct legal relationship to either the Corporation for Public Broadcasting (CPB), which is an independent agency supported by congressional funding, or with the Public Broadcasting Service (PBS) or National Public Radio (NPR), which were established by CPB as national television and radio distribution systems, respectively.

Educational/Public Radio   Twenty FM channels between 88 and 92 MHz are reserved for noncommercial radio stations. About 80% of all noncommercial radio stations are licensed to colleges and universities, with the remainder licensed to state or local school systems and to nonprofit educational citizen organizations. All noncommercial radio stations, whether called public or educational, are licensed by the FCC. CPB and NPR often give the impression that only their members are public radio stations; however, they have restricted membership to the wealthier stations, those able to afford five full-time paid employees, among other requirements, and their membership consists of only about 400 of the more than 1,400 noncommercial radio stations licensed by the FCC. The remainder are denied the use of the public tax monies for public radio that go to CPB and NPR.

While there are some 25 AM stations licensed to nonprofit educational entities and operating noncommercially, they are not officially noncommercial stations. They are the residue of the more than 170 educational licensees in the 1920s, when radio began. Most of these went off the air when the Federal Radio Commission reallocated frequency assignments in 1927, giving many of the better frequencies occupied by educational stations to commercial interests and eliminating a number of educational stations entirely. There are no reserved allocations in the AM band.

Carrier-current   Many colleges and universities have what are called carrier-current radio stations. Such a station operates on extremely low power on a regular radio frequency. The signal is carried through metal conduits, such as pipes, to dormitory rooms and other selected receiving points on the campus where it can be received on the regular AM radio dial. Because the signal literally does not go out over the air, carrier-current radio is not regulated by the FCC. However, where a carrier-current station uses power of a magnitude that causes the AM signal to spill into the airwaves beyond the confines of the campus and to interfere with a regularly licensed operation, the FCC will step in.

Educational/public television   In 1952 the FCC reserved 242 channels for noncommercial television use only; 80 were VHF and 162 were UHF. Over the years more reserved channels were added; in 1966 alone the FCC designating 615 UHF channels for educational TV. In early 1991 there were 125 VHF and 229 UHF educational TV stations on the air. About one-third of all educational/public television stations are licensed to colleges and universities, about one-third to state and local educational systems, and about one-third to nonprofit educational organizations operated by citizen groups. A number of states have public television station networks, many of the stations in some systems serving as relay stations for the signal from the main studio.

ITFS   The Instructional Television Fixed Service (ITFS) was originally established by the FCC as special frequencies in the 2500–2690 MHz band for the exclusive use of nonprofit educational institutions or organizations for the primary purpose of formal instructional programming. ITFS has gradually been reregulated by the FCC to give more and more of those frequencies to private commercial use. ITFS is considered a point-to-point microwave, operating much like a broadcast station, but requiring a special receiving antenna and down-converter for reception. It is used for safety, health, and professional updating purposes as well as for formal instruction on all academic levels. ITFS is handled by the Distribution Services Branch.

Closed-circuit television   Many schools and other institutions, such as hospitals, have closed-circuit systems that connect a studio or other source of programming to individual sets via wire. These systems do not use the airwaves, and therefore do not fall under FCC regulation.

Auxiliary services   When covering on-the-spot events, whether news, sports, conventions, openings of supermarkets, or other *remote* happenings, broadcasters use portable transmitters to relay the visual material, aural material, or both back to the station live. Spectrum space is needed for these radio and television transmitters. In addition, stations use special transmitters to send the signals from the studios to the transmitters and to relay programs between broadcast stations. These studio-transmitter links (STLs) and intercity relay stations are also regulated and specifically licensed. An Auxiliary Services Branch in the Audio Services Division and a Distribution Services Branch in the Video Services Division are responsible for auxiliary services and, to varying degrees, for other aspects of radio and television that do not fall directly under AM or FM, broadcast television, low power television, or cable.

Stereophonic Services   AM, FM, and TV stations are all able to transmit stereophonic sound programs. Stereophonic systems must be approved by the FCC. FM has transmitted in stereo for many years, and in recent years more and more TV stations have done so. The FCC authorized AM stereo in 1982, but did not designate any one system, leaving it to the marketplace to determine which of the five approved alternative technical standards would win out. Because of the incompatibility of the various systems, by the early 1990s relatively few AM stations had yet converted to stereo.

Advanced/High Definition/Enhanced Definition Television These terms—ATV, HDTV, and EDTV—frequently are used synonymously. They are not synonymous. ATV refers to any system of distributing television programming that results in better video and audio quality than that of the U.S. NTSC standard of 525 lines, recommended by the National Television Standards Committee when television was first authorized in the United States, giving the United States the poorest quality of the three TV distribution standards subsequently in use throughout the world. HDTV offers about twice the number of lines, with picture quality comparable to that of 35mm film and audio quality similar to that of compact discs. EDTV refers to systems that are an improvement over NTSC, but not as good as HDTV.

In the late 1980s, as other countries—principally Japan—made great strides in the introduction of HDTV, the FCC issued a Notice of Inquiry (NOI) and then a Further NOI, in which it sought to determine whether and what kind of an ATV system the United States should establish. At the beginning of the 1990s the FCC had not yet made a final decision, but it had determined it preferred an HDTV system, rather than EDTV, that can operate in the present spectrum allowed for broadcast television and that is compatible with the NTSC system—that is, permitting existing sets to receive the new signal in the 525-line mode while the public gradually switched over to the advanced system. This is the approach used when color television was authorized in 1953. Much of the work done on developing a new system is through an industry-wide Advisory Committee on Advanced Television Service set up by the FCC in 1987. While a number of FCC offices are working on ATV, because the Commission proposes to use it for broadcast purposes only, the Mass Media Bureau is principally involved.

Cable The FCC first established rules relating to cable in 1965, and in 1972 published a comprehensive set of regulations governing this still relatively new distribution system. The Cable Communications Act of 1984 deregulated most cable rules, including those relating to subscriber rates, required programming, and fees to municipalities. The FCC was left with virtually no jurisdiction over cable. At one time the FCC had a Cable Bureau; presently those cable regulations that remain are implemented by the Cable Television Branch of the Video Services Division. Among the remaining areas of FCC jurisdiction are the registration of each community system prior to beginning operations; offering subscribers A-B switch installation to permit off-the-air as well as cable reception; carriage of TV broadcast programs in full, without alteration or deletion, unless the FCC grants a waiver to permit commercial inserts; nonduplication of network programs; syndicated program exclusivity for broadcast stations for non-network programming; possible fines, imprisonment, or both for conviction of carrying obscene material; licensing for receive-only earth stations for satellite-delivered pay cable (in 1991 there were about 10,000 such licensees and an estimated two million receive stations operating without a license); conformity with the FCC rules on equal employment opportunity in regard to race, color, religion, national origin, age, or gender; and licensing for any frequency use by cable, including cable television relay service (CARS) stations. The FCC may impose fines on cable systems violating its rules. In 1991 Congress was working on legislation to reregulate cable, including the restoration of some FCC jurisdiction.

PRIVATE RADIO BUREAU

The Private Radio Bureau regulates services meeting the private communications needs of businesses, individuals, nonprofit organizations, and state and local government. These include police, fire, and safety systems on land, sea, and air, and private communications within business and industry on land, such as transportation, oil, forestry, agriculture, and real estate. Private stations and services far outnumber the more than 20,000 stations of various kinds licensed by the Mass Media Bureau. In fact, private radio services are used by literally millions. There are six major service categories for private radio: maritime, aviation, private land mobile, operational fixed microwave, personal radio, and amateur radio. We are all familiar with the large group of amateur radio operators called hams. The many professional operators of private radio include police officers, firefighters, farmers, factory workers, truck and taxicab drivers, airplane pilots, ship officers, health and safety workers, and dispatchers of all kinds. Most of us encounter several uses of private radio every day, even though many of us may not be aware of it.

The Private Radio Bureau is divided into an Administration and Management Staff, which performs the same management functions for its Bureau as the comparable staff does for the Mass Media Bureau, and three Divisions:

1. Land Mobile and Microwave Division, with a Policy and Planning Branch, a Rules Branch, and a Compliance Branch;
2. Special Services Division, with an International Liaison Staff, an Aviation and Marine Branch, and a Personal Radio Branch;
3. Licensing Division, responsible for the millions of private radio users, with a large number of support Branches and Sections to enable it to do its job. These include the Land Mobile Branch, Microwave Branch, Support Services Branch, Systems and Procedures Branch, Special Services Branch, Consumer Assistance Branch, Mail and Fee Branch, Aircraft and Ship Section, Aviation Ground and Marine Coast Section, General Radio Section, Special and General Facilities and Technical Sections, and a number of sections that deal solely with the process of licensing and keeping records (see Figure 7).

Private Radio Terms   Common terms used in private radio, useful in understanding the nature of the Private Radio Bureau's regulation of the key services under its jurisdiction, are:

base station: a station in the land mobile or personal service at a specific location or base, which is used to carry on communications with mobile stations;

fixed station: a station at a specified or fixed location, which is used to carry on communications with other fixed stations;

land station: a station in a mobile service not intended to be used while in motion; for example, base stations, remaining at specified locations, are land stations;

mobile station: a station that may be used while in motion, as in a vehicle or as a portable unit;

mobile relay stations: a base station used to retransmit communications to and from mobile stations.

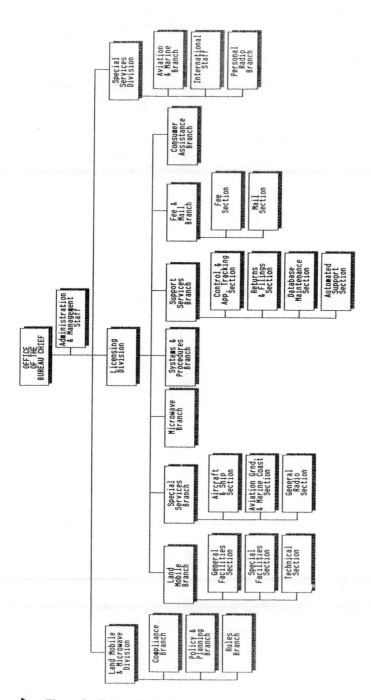

▶ *Figure 7   Private Radio Bureau organizational chart.*

Special Services Division

This division is responsible for all FCC regulatory, compliance, and international matters relative to the Aviation, Marine, and Personal Radio services, including rule interpretation, interagency and public liaison, international representation, issue analyses, and rule-making proposals.

The *Aviation and Marine Branch* administers the compulsory ship radio safety requirements, including policy and rule-making interpretations and recommendations, liaison with government and industry groups, and investigation of possible violations and implementation of sanctions.

The *Personal Radio Branch* has identical duties for the Amateur Radio Operators Qualification Program, plus the preparation of syllabi and requirements for volunteer examinations and the overseeing of the volunteer program.

The *International Liaison Staff* coordinates the Commission's international telecommunications activities in private radio matters, including appropriate studies, preparation of documents for international meetings, proposals for treaty changes, and proposed rule-making to conform to international agreements.

Land Mobile and Microwave Division

This division is responsible for all applications, rule-making, and regulatory and compliance activities relating to the Private Land Mobile and the Operational Fixed Microwave Radio services. It interprets the rules, prepares legislation, analyzes legal, technical, and economic aspects of the services, does long-range planning regarding future regulations and policies, and prepares materials for other Commission offices, and for conferences and Congressional testimony.

The *Policy and Planning Branch* studies and recommends policies and rules, provides economic data and legal and technical information to other offices, and evaluates the long-range potential contributions of private radio to the country.

The *Rules Branch* evaluates current Rules and Regulations, studies legal and engineering trends, including interference standards and frequency assignment criteria, and develops proposed rule-making and legislative proposals. It evaluates new technologies, maintains information liaison with the public and the industry on land mobile and operational fixed stations, evaluates rule-making petitions, and negotiates bilateral agreements at international conferences regarding its regulated services.

The *Compliance Branch* administers the Bureau's enforcement program by investigating all violations cases with the Field Operations Bureau, rendering initial penalty decisions, and preparing hearing and enforcement proceedings, including those for referral to the Department of Justice or a U.S. Attorney. It maintains liaison with public and industry groups and tries to provide direct assistance in problem solving, especially those relating to interference.

Licensing Division

This division is responsible for authorizing service in all private radio services, and establishes guidelines, processes applications, offers policy and rule interpretations and other assistance related to the processing, and maintains liaison with national and international groups on licensing matters.

The *Systems and Procedures Branch* is principally responsible for the automated and manual application processing systems for all private radio services offices, evaluates the efficiency of processing systems, and designs and modifies the license forms. It functions as the division's managing team, including personnel evaluation.

The *Land Mobile Branch* authorizes and administers applications and licenses for the Private Land Mobile radio services, including Specialized Mobile Radio System, Private Carrier Paging, Industrial, Public Safety, and Land Transportation Radio. It recommends policies and rule-making pertaining to these services. The branch's *General Facilities Section* deals with new, modified, and renewal authorizations for frequencies outside of the normal FCC selection or criteria, principally in the proximities of the Canadian and Mexican borders. It coordinates with other agencies and policies, including the National Environmental Policy Act, the National Radio Astronomy Observatory, and the International Radio Advisory Committee (IRAC). Its *Special Facilities Section* is responsible for applications for so-called complex private land mobile radio systems, performing similar functions for its area as does the General Facilities Section. It handles such services as the Emergency Medical, Quiet Zone, Radiolocation, and Travelers Information stations. The branch's *Technical Section* provides technical advice and rule interpretations to internal and external offices and the public, resolves technical problems on applications and other frequency utilization documents, handles rule waiver requests for technical relief, conducts studies on frequency assignments, spectrum usage, radiowave propagation and other technical matters for purposes of rule-making recommendations, handles interference complaints, and conducts negotiations with foreign governments on private radio matters. It may take action on applications for systems not included in the FCC Rules and Regulations.

The *Microwave Branch* authorizes and administers the Private Operational Fixed Microwave Service, performing the same duties in regard to studies, rule recommendations and interpretations, liaison, complaints, and similar matters as do the other branches in respect to their services. The Microwave Branch also maintains an automated microwave database.

The *Special Services Branch* has similar functions to the offices above, in its authorization and administering responsibilities for Aviation and Maritime Radio, Alaska Public Fixed, Amateur and General Mobile Radio, and Citizens Band and Radio Control services. These functions are implemented by the branch's *Aircraft and Ship Section*, which deals with Aviation Aircraft and Marine Ship services, its *Aviation Ground and Marine Coast Section*, which handles applications for radio stations in the Aviation Ground, Marine Coast, and Alaska Public Fixed Station, and its *General Radio Section*, with responsibility for Amateur and General Mobile Radio services.

The *Support Service Branch* handles mail fee collecting and refund requests, application files, computer support, data collection, and other quality control areas. It also processes renewals for Land Mobile Radio and Restricted Telephone Operator Permits, and applications for Citizen Band and Radio Control.

The branch's *Applications Tracking Unit* operates a tracking system for microwave, ground, and coast stations, and maintains a public file on these operations. Its *Returns and Files Section* returns defective applications for correction and maintains files on applications processed. Its *Database Maintenance Unit* processes renewals for Land Mobile, Aviation, and Marine services, maintains detailed statistical information on its regulated services through an automated database file, and prescreens applications to determine problems that need special attention of other offices. Its *Automated Support Section* works with the programming and operating of existing hardware, maintains a tape library of software, and collects statistical data.

The *Consumer Assistance Branch* handles day-to-day inquiries from the public, industry, and Congress on applications and procedures, and handles Freedom of Information Act requests.

The *Fee and Mail Branch* opens, sorts, and distributes mail and logs in applications with fees for processing. The branch's *Mail Section* opens, sorts, and distributes applications to appropriate offices, grants or returns applications for restricted operator permits (which may be obtained by application without an examination), and mails out licenses. Its *Fee Section* maintains an accounting system for all fees collected, coordinates the fee collection activities, and makes recommendations concerning the fee collection process.

Many services regulated by the Private Radio Bureau are either especially known to the public or play key roles in the overall communications patterns in the country. Further discussion and clarification of some of these follow.

Key Regulatory Areas

Maritime Services   Although both the maritime and aviation services provide communications to the public under some circumstances, they are intended primarily for the safety and the private use of persons engaged in maritime or aeronautical operations. The Maritime Radio Services include the Maritime Mobile Service, the Maritime Mobile-satellite Service, the Maritime Radiodetermination Service (essentially radar stations for navigation), and stations in the Fixed Service. The two major classifications of maritime radio services are stations on shipboard and stations on land. Depending on size, purpose and destination, some shipboard stations are mandatory, others optional. Shipboard stations may use telephone or telegraph, may have portable radiotelephones, may include survival craft stations used in lifeboats or rafts, and may be connected via satellite for ship-to-shore communication.

Land stations provide the links between ships and shore, and include coast stations as the most direct communication connections, operational fixed stations for repeater or relay purposes, radar stations for locating and tracking vessels, radiobeacons (such as lighthouses) for location purposes, and maritime support stations for training and equipment testing-purposes.

The FCC regulates these services for all U.S. registry ships, including those that sail in international waters, and for all ship operations in U.S. waters. Rules governing requirements and licensing in domestic waters are developed in coordination with the U.S. Coast Guard.

Aviation Services   Most radio stations used in aviation are part of the Aeronautical Mobile Service, which includes both airborne and land stations. The Private Radio Bureau licenses eleven different types of such stations. They include airborne stations, those used in flight, whether on an airplane, helicopter, or even a hot-air balloon. Messages to and from airborne stations are those limited to safe flight operations. Advisory, or unicom stations, are on land and advise pilots about local airport conditions, but are not used to control aircraft in flight. The latter is accomplished through airdrome control stations, which provide communications on landing and taking off, and en route stations, which are used for operational control and flight management. Multicom stations permit communications between private aircraft and a ground facility for temporary, seasonal, or emergency activities, such as forest firefighting, crop dusting, and livestock herding. Utility mobile stations are installed in vehicles that provide maintenance, fire and crash protection, freight handling, and other support activities. In addition, there are search and rescue stations, automatic weather observation stations, and flight test and instructional stations.

Another type of station, the aeronautical radionavigation service, is operated mostly by the Federal Aviation Administration (FAA), but the FCC does license some of these stations where a service is needed that the FAA does not provide. The FCC authorization in such instances requires operators to comply with FAA standards. The FCC cooperates with the FAA in all aviation service regulation, with applications for operations on U.S. territory having to meet both FCC and FAA requirements.

Private Land Mobile Services   Private land mobile radio encompasses the bulk of FCC licensees and many fields of endeavor such as newspapers, public utilities, transportation industries, farmers and foresters, police departments, manufacturers, and motion picture producers. Mobile service was defined by the International Radiotelegraph Convention of 1927 as that "carried on between mobile stations and land stations, and by mobile stations communicating with one another." The Private Radio Bureau licenses 20 specific services in five major service categories.

Public safety radio services are used for safety and emergency purposes by local government, police, fire, forestry-conservation, and highway maintenance users. Local government sometimes equips different agencies with the same frequency to coordinate special activities, from utility repairs to disasters such as floods. Police use is the oldest public safety activity, and includes voice, data, teletype, and video. Virtually all police vehicles and even cops-on-the-beat, as well as police stations, are equipped with intercommunication devices. The fire radio service provides communication between headquarters and fire vehicles, and coordinates the work of fire fighters at the scene of a blaze. While principally licensing fire stations under local government control, the FCC also accepts radio applications from volunteer fire companies. Forestry-conservation services facilitate communication networks among fire lookout towers and ground crews and for game law enforcement, flood and erosion control, and insect and disease protection. The highway maintenance service offers emergency and routine communication links to state and local governments for purposes of road safety, snowplowing, debris clearing, road repair, and similar efforts.

Special emergency radio service is oriented to protection of life and property, and includes the communication needs of hospitals, ambulances, disaster relief organizations, school buses, beach patrols, veterinarians, and persons and organizations in isolated areas.

Industrial radio services is a catch-all area for the industrial uses of radio that developed over many years. Presently, there are nine categories of licensing. The

Power Radio Service is for utilities like electric, gas, and water companies in connection with producing, maintaining, and distributing their services. The Petroleum Radio Service is used by oil and natural gas industries for exploration, production, refining, and pipeline distribution. Some frequencies are available for signals that determine the characteristics of the earth in relation to potential oil fields. The Forest Products Radio Service goes beyond safety and forest protection and is used for communications in logging, publishing, and manufacturing lumber, paper, and similar products. The Motion Picture Radio Service is used for communicating either on location or in the studio. The Relay Press Service is used on location by reporters and photographers covering assignments. The Special Industrial Radio Service provides communication for farming, ranching, road and bridge construction, mining, distributing chemical products, drilling wells, and distributing fuel or ready-mixed concrete. The Business Radio Service is a miscellaneous category covering communications by commercial enterprises, nonprofit organizations, churches, or medical groups. Its broad areas of inclusion make it the most widely used land mobile service. The Manufacturers Radio Service offers radio communication for production, safety, logistics, and the handling of machines and materials at plants, factories, mills, and shipyards. The Telephone Maintenance Radio Service is for telephone and telegraph companies use in connection with construction, repairs, and maintenance of efficient operation of their communication systems.

Radiolocation service refers to the use of radio waves to determine an object's speed, distance, direction, or position. Radar is an example. People engaged in commercial, industrial, scientific, educational, or governmental activities for these kinds of study—with the exception of navigation—must obtain FCC authorization.

Land transportation radio services provides communication authorization for different types of transportation, including railroads, urban transit (such as subways), taxicabs, intercity buses, highway trucks, and automobile emergency vehicles. The Motor Carrier Service is used by bus and trucking companies carrying passengers and freight, for communication between terminals and vehicles. The Railroad Service covers all communication on the same train or between trains, between terminals and trains, and in activities in railroad yards. The Taxicab Service links company dispatchers with drivers. The Automobile Emergency Service is used principally by auto clubs and garages for the servicing of disabled vehicles.

The FCC permits licensees in private land mobile services to share their stations with others under certain circumstances. Under multiple-licensing different users may be licensed independently for a station they all share; some single-licensed stations can be used on a cooperative basis. While private radio users are licensed to operate their facilities only for their own purposes, there are two exceptions which allow private licensees to offer their radio services for hire: Specialized Mobile Radio System (SMRS) offers two-way mobile dispatch services for hire, and Private Carrier Paging Systems (PCPS) offers a one-way paging service.

Private Operational Fixed Microwave Service   Microwave communication consists of very short waves in the upper range of the radio spectrum. It is used principally for point-to-point communication, rather than, like broadcasting, reaching out to a mass audience. Operational fixed stations may be used for commercial, industrial, or safety purposes, sending signals from a headquarters to a specified remote site, or from a central location to a number of remote sites. Some systems carry long-distance communications through microwave links installed along a pipeline, railroad, or highway. Operational fixed is a convenient, cost-effective alternative to cable for the transmission of voice, data, and video signals to a specified receiving point.

Operational fixed microwave operates identically to ITFS, discussed under the Mass Media Bureau. The signals are sent out over the air and are picked up by a special receiving antenna and down converter, converting the extremely high-frequency signal to one tunable on an unused channel on an ordinary television set. The material received is then distributed by wire to different buildings in the industrial complex or to different offices or rooms in the same building.

Multichannel Multipoint Distribution Systems (MMDS) use operational fixed microwave to provide commercial material, such as teleconferences for businesses, special programs, including such nonbroadcast materials as local events and porno movies to hotel TV sets, and entertainment programs to multiple receiver sites, such as hospitals, apartment complexes, hotels, and nursing homes. MMDS has over the years received from the FCC some of the channels originally reserved for educational institutions in the ITFS band. Some MMDS systems, with multiple channels in a given city, have attempted to become competitive to cable, and are sometimes called wireless cable.

Personal Radio Services   Almost anyone can use the personal radio services, which provide low-cost short-range communications for personal or business purposes. No license is required for the Citizens Band (CB) and the Radio Control (R/C) services. CB, a 40-channel two-way radio service, and R/C, one-way non-voice operation of devices by remote control, may not be operated by a foreign government or representative and must use an FCC type-accepted transmitter. General Mobile Radio service (GMRS) is a 16-channel, two-way FM voice system that requires a license, with the FCC designating the channel to be used. Each station operator must have permission from the legal licensee.

Amateur Radio Service   A synonym is *ham* radio. There are no age restrictions or equipment standard requirements for operators. However, ham stations may be used only for personal and not for business purposes. Ham operators may transmit voice, telegraphy, teleprinting, television, and facsimile. They may communicate with other amateur radio operators anywhere in the world. A ham operator must obtain a license from the Private Radio Bureau. Tests must be passed for one of five classes of amateur operator licenses, each higher class conveying additional operational privilages to the amateur, and requiring increasing skill and knowledge in such areas as international Morse code, radio wave propagation characteristics, electrical principles, knowledge of the FCC rules for amateur stations, and signals, emissions, antennas, and feed lines.

COMMON CARRIER BUREAU

Common carrier refers to communication companies that hire out their specific authorized services to the public. Telephone, telegraph and some satellites are examples of common carriers. Local telephone companies are common carriers inasmuch as they must offer telephone service on an equal basis (that is, at the same rate to anyone who can pay) in homes, businesses, or at public pay phones.

The FCC's Common Carrier Bureau regulates interstate and foreign common carriers operating within and from the United States, whether by wire, radio, cable, or satellite. *Intra*state communication is not subject to federal government regulation, but is under the jurisdiction of state utility commissions. Common Carrier's authority includes licensing of radiotelephone circuits, assigning frequencies for their operation, considering applications for construction of new facilities, approving discontinuance or reduction of service, acting on applications for mergers, and regulating rates for interstate telephone and telegraph services and for services between the United States, foreign and overseas points, and ships at sea.

The Bureau also is authorized to prescribe the forms of records and accounts kept by common carriers, and has established uniform systems that include original cost accounting, continuity property records, pension cost records, and depreciation records. It takes part in policy, adjudicatory, and rule-making proceedings, works with state regulatory commissions and the National Association of Regulatory Utility Commissioners (NARUC), and participates in international conferences concerning common carrier matters.

More than 1400 privately owned telephone companies provide local telephone service in the United States, including 22 Bell Operating Companies (BOCs) that were formerly part of a nationwide integrated Bell system. Following the FCC's AT&T divestiture ruling in 1982, seven regional holding companies replaced "Ma Bell's" national monopoly and currently serve about 85% of consumer common-carrier telecommunications needs in the country. The remaining 15% of the population is served by a small number of large independent telephone companies and by about 1400 small local companies. In addition, a number of firms provide long distance service in the United States and to foreign countries. At the beginning of the 1990s some 300 companies were attempting to compete with AT&T as a result of the divestiture ruling's provision for equal access to long-distance trunk connections. The Common Carrier Bureau is divided into nine different Divisions to implement FCC rules and policies. These divisions cover many of the same technical types of communication services under the jurisdiction of the Mass Media and Private Radio Bureaus, but in terms of their common carrier applications only.

1. The *Domestic Facilities Division* has a Domestic Radio Branch, a Domestic Services Branch, and a Satellite Radio Branch;
2. The *Mobile Services Division* contains a Legal Branch, Cellular Radio Branch, Public Mobile Radio branch, and a service office, the Public Reference and Information Branch;
3. The *Accounts and Audits Division* enforces the Common Carrier Bureau's jurisdiction over licensee accounting through its Accounting Systems, Audits, Cost Analysis, Depreciation Rates, and Legal Branches;
4. The *Enforcement Division* deals with alleged violations of the rules through its Formal Complaints and Investigations Branch and Informal Complaints and Public Inquiries Branch;

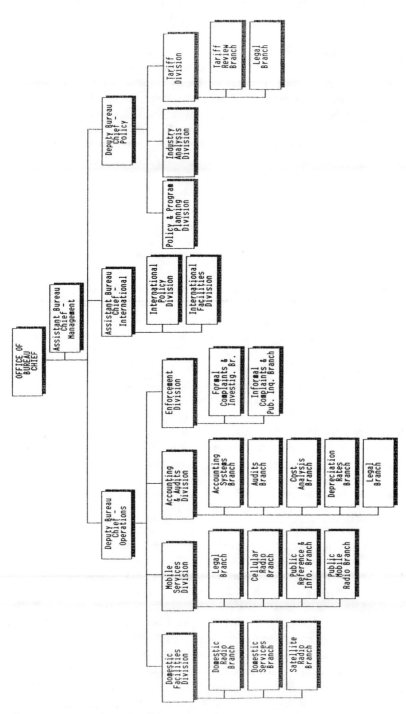

▶ *Figure 8 Common Carrier Bureau organizational chart.*

The four Divisions above report to a Deputy Bureau Chief/Operations.

5. and 6. The *International Policy* and *International Facilities* Divisions are under an Assistant Bureau Chief/International.
7., 8., and 9. A Deputy Bureau Chief/Policy supervises the *Policy and Program Planning*, *Industry Analysis*, and *Tariff* Divisions; the latter contains Tariff Review and Legal Branches.

The Bureau also has an Assistant Bureau Chief/Management, whose duties are the same for the Common Carrier Bureau as are the counterpart positions in the Mass Media and Private Radio Bureaus (see Figure 8).

Domestic Facilities Division

This office develops, recommends and implements policies, rules, standards, procedures, and forms for the authorization and regulation of common-carrier domestic wireline, cable and microwave facilities, and national and international satellite and radio. It issues licenses, coordinates frequency assignments, and participates in adjudicatory hearings.

The *Domestic Radio Branch* develops and implements rules and procedures for the authorization and regulation of common carrier point-to-point microwave, local television transmission, and multipoint distribution services.

The *Domestic Services Branch* is responsible for rules and authorization procedures for transmission facilities used by interstate common carriers.

The *Satellite Radio Branch* deals with common carrier satellite radio stations.

Mobile Services Division

This office authorizes and regulates domestic common carrier public mobile services that include Public Land Mobile, Rural Radio, Offshore Radio, Air-Ground Radiotelephone, and Public Cellular Telecommunications.

The *Legal Branch* is responsible for developing and interpreting nontechnical aspects of the Rules and Regulations and monitoring their compliance by licensees. It also reviews the legal portions of cellular radio and public mobile radio applications.

The *Public Mobile Radio Branch* administers the authorization, regulation, and compliance of all public mobile radio services except cellular.

The *Cellular Radio Branch* handles the authorization and regulation of domestic public cellular telecommunications. Cellular radio is discussed further later in this section.

The *Public Reference and Information Branch* accepts, logs, issues public notices, and distributes for processing all mobile service applications. It maintains public reference files on applications and grants and provides information and handles public inquiries.

Accounting and Audits Division

This Division monitors the FCC requirements of common carriers for reporting and for preservation of records. It does field audits and investigations of carriers' financial and operating practices, procedures, and records. It recommends depreciation rates and financial- and cost-modeling systems.

The *Accounting Systems Branch* establishes and periodically revises the uniform accounting and reporting systems, develops short- and long-range recommendations for rule-making in this area, maintains liaison with other Commission offices, federal agencies, the industry, and the public, and maintains oversight on pension plans.

The *Cost Analysis Branch* develops studies, proposed rules, model systems, Cost Allocation Manuals, and automated analyzing systems for jurisdictional separation, access costs, rate bases, and tariff reviews. It works closely with the Federal-State Joint Board.

The *Audits Branch* establishes audit programs and conducts field audits, with follow-up evaluations and recommendations to correct deficiencies. It studies common carriers' life and net salvage factors, depreciation reserve requirements, book reserves, and reserve deficiencies and surpluses. It reviews the Bureau's audit program, and sometimes conducts internal audits of the FCC.

The *Legal Branch* provides legal guidance to the division on the requirements of the Communications Act, FCC procedures and rules, and the Freedom of Information Act. It acts on petitions for waivers of the rules, and analyzes court decisions and other legal requirements imposed on common carriers to determine compliance actions by the division.

Enforcement Division

This division analyzes complaints, conducts investigations, and enforces compliance with requirements concerning reasonable rates, terms, and conditions for cable pole attachments. It also handles requests under the Freedom of Information, Privacy, and Regulatory Flexibility Acts.

The *Formal Complaints and Investigations Branch* investigates and resolves complaints from the public, federal agencies, and common carriers concerning rule violations, including those related to cable television pole attachments and common carrier mergers. It participates in litigation matters, arranges audits, prepares cases for hearing, and participates in oral arguments. The *Informal Complaints and Public Inquiries Branch* handles all kinds of informal complaints and inquiries, including those from the public, Congress, and the industry, and recommends rule-making based on complaint trends. It publishes information bulletins from time to time designed to inform the public on matters about which it has received many inquiries.

Policy and Program Planning Division

This division develops long-range objectives and proposes new policies and rules based on a study of legal, economic, financial, and technological factors. It seeks greater competition in the industry (for example, the divestiture of AT&T) and greater efficiency in the regulatory process. It encourages new firms, services, and equipment, and is concerned with rates, ownership, and interconnection.

Industry Analysis Division

This office statistically monitors the impact of FCC policy on the industry, and establishes reporting requirements and processes for industry financial data. It studies the effects of the AT&T divestiture on the nationwide telecommunications network structure and operation. It provides technical advice and recommends rule-making to the Commission based on its studies, especially its economic analyses. It measures the impact of the FCC's international telecommunications policies on international communications. It administers the FCC's EEO program for common carriers.

Tariff Division

This division administers the tariff provisions of the Communications Act, seeing to it that the charges, classifications, regulations, and practices of common carriers are just, reasonable, and not discriminatory.

The *Tariff Review Branch* reviews carriers' tariff schedules and revenue requirements for their compliance with FCC rules, and initiates enforcement actions when necessary. It recommends rule-making regarding tariff regulations and procedures, including those designed to simplify tariffs and expedite action on their filings.

The *Legal Branch* makes recommendations regarding legal requirements imposed on carriers regarding tariffs, handles waiver requests, prepares the documents involved in formal proceedings on tariff matters, and recommends formal investigations on carrier practices.

Assistant Bureau Chief/International

This office develops and coordinates policies, rules, and procedures for international communications planning, facilities, and authorizations, maintains liaison with the Department of State and other U.S. agencies and organizations and with international groups, and plays prominent leadership roles at international conferences. This position also has an oversight role in relation to COMSAT's participation in INTELSAT, and authorizes facilities and services of international communication carriers including cable, radio, and earth stations. The Assistant Bureau Chief directs two divisions that carry out these responsibilities.

International Policy Division   The International Policy Division develops recommendations for international common carrier regulation, interprets the rules, and seeks more efficient regulation and procedures. It develops positions and proposals and is part of U.S. delegations at international meetings, including those of the International Telegraph and Telephone Consultive Committee, International Radio Consultive Committee, Inter-American Telecommunications Conference, and INTELSAT and INMARSAT.

International Facilities Division   develops policies and standards and authorizes new services and facilities, including alternative international common-carrier systems utilizing submarine cable, satellites, high-frequency radio, and other distribution means. It processes applications for international voice, data, and record services and monitors rules compliance by licensees.

Key Regularity Areas

While the Common Carrier Bureau, as noted in the description of its various offices, has a wide range of responsibilities in many communication areas, perhaps the most significant in terms of impact on and/or awareness of the public are (1) telephone, especially AT&T divestiture, competition in providing users' equipment, rates, and additional long distance services; (2) cellular mobile radio; (3) domestic satellites; and (4) computer networks. A brief overview of its activities in these areas will provide pertinent examples of the Common Carrier Bureau's key work.

Telephone   The authority of the FCC in regard to telephone and telegraph was clearly evidenced in the divestiture or "breakup" of the AT&T monopoly on equipment and services in the United States. The Common Carrier Bureau sees to it that the Bell Operating Companies do not violate the prohibitions on providing long-distance services outside of designated geographical areas, providing information services, and manufacturing equipment. The Bureau also monitors the requirement that BOCs provide local exchange access to all long-distance carriers on a "basis that is equal in type, quality, and price to that provided to AT&T."

The FCC also gave consumers the option to buy their own telephones for use with AT&T services, rather than being obligated to rent the instrument from AT&T, as long as the equipment meets FCC standards. The FCC monitors the operations of the myriad long-distance companies that have entered the field, such as MCI and Sprint. The Common Carrier Bureau continues to monitor and may require adjustments in rates, and must approve of any rate change.

Cellular Radio   The FCC's authorization of cellular radio greatly increased mobile communications for the public. Fifty MHz of electronic spectrum (824–849 MHz and 869–894 MHz) are available to two competing systems in each market, including local telephone companies, with the companies selected by lottery in the top 30 markets, subject to legal and technical review. The cellular system is further divided into smaller geographic areas in each market, linked to the local telephone company's central office, so that local, long-distance, and international calls can be completed from a cellular phone without requiring an operator.

By the spring of 1990 the FCC had issued construction permits and licenses for cellular radio service in all of the 306 Metropolitan Statistical Areas (MSAs) in the United States, serving about 75% of the population. In addition, by the same date, the FCC had granted 322 construction permits for rural areas. The FCC provides protection for cellular licensees within their metropolitan or rural area for 5 years, prohibiting any other company from filing for a license. Users of cellular phones are increasing by the millions each year.

Domestic Satellites   A satellite is a communications relay station that orbits the earth and receives and sends signals to earth stations, allowing the earth stations to communicate with each other at long distances. Satellites are used to transmit telephone, television, radio, and data signals originated by common carriers, broadcasters, and cable distributors. Satellite operators are subject to the same regulatory controls by the FCC as are other common carriers. The FCC licenses all satellite carriers, oversees competition in the procurement of equipment, approves the technical characteristics of the system, and authorizes terminal stations in the United States.

Domestic satellites for communication have been licensed under an *open entry* policy since 1966. Carriers are allowed to sell transponders (the antenna-like part of the satellite that receives signals from the earth, translates and amplifies them, and then retransmits them) to non-common carriers, and non-common carriers are allowed to own and operate their own satellites.

Satellite piracy and scrambling are concerns of the FCC. Section 705 of the Communications Act protects the privacy of pay-TV services, such as cable, MMDS, subscription TV, and DBS. Wholesale piracy of satellite signals by satellite-receiving dish owners who have not paid fees to the originating company has resulted in signals being scrambled by those companies. It is illegal to intercept and unscramble such signals (HBO, for example) without authorization, which usually requires the paying of a fee. Because piracy of satellite signals is a violation of the U.S. Criminal Code, such theft is under the jurisdiction of the Department of Justice. The FCC, however, is involved in direct investigations of signal theft and works closely with the Department of Justice in prevention and prosecution.

Computer Networks   With the growth of computer networks the FCC assumed regulatory jurisdiction over common carrier computer services. The Commission divides common carriers into *dominant* and *non-dominant* categories, the former referring to those companies, such as AT&T, that have strong market control and can affect the cost of service by restricting output, and the latter to those carriers without much market power. Dominant carriers are required to obtain a license and file tariffs, which the FCC can accept, suspend, or reject. Non-dominant carriers do not need licenses and are not required to file tariffs for domestic computer network services, giving them the flexibility of providing users lower rates and better services. However, they must obtain a license for overseas services. The FCC also has rules designed to prevent dominant carriers from unfairly competing against non-dominant carriers through the former's control of basic communication services. For example, AT&T and BOCs must provide "comparably efficient interconnections" to the computer network for the telephone companies' competitors, and must not use their greater information about customers to unfairly compete against non-dominants.

FIELD OPERATIONS BUREAU

The Field Operations Bureau is the FCC's technical watchdog. It detects violations of radio regulation, monitors transmissions, inspects stations, investigates complaints of radio interference, and issues notices of violations. It also examines and licenses radio operators, processes applications for painting, lighting and placing antenna towers, and furnishes direction-finding aids for ships and aircraft in distress. The Bureau maintains, as of the spring of 1990, 36 field offices throughout the United States. The heads of these offices report to six regional directors located in Boston, Atlanta, Chicago, Kansas City, Seattle, and San Francisco, who in turn report directly to the Field Operations Bureau Chief in Washington.

The Bureau has an Administrative Office and a Program Development and

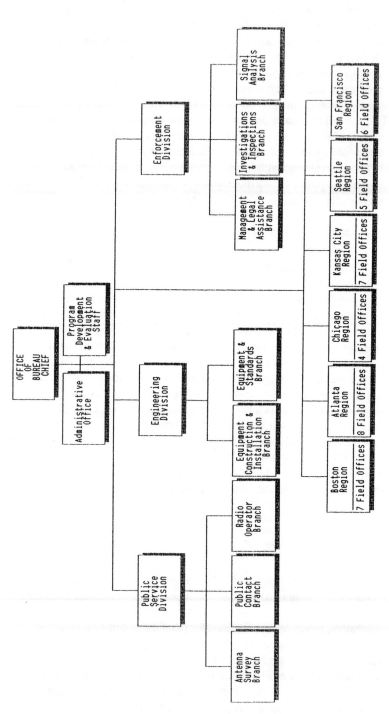

▶ *Figure 9   Field Operations Bureau organizational chart.*

Evaluation Staff, and three divisions dealing, respectively, with Public Service, Engineering, and Enforcement:

1. The Public Service Division has Antenna Survey, Radio Operator, and Public Contact Branches;
2. The Engineering Division contains an Equipment Construction and Installation Branch and an Equipment and Standards Branch;
3. The Enforcement Division has Management and Legal Assistance, Investigations and Inspections, and Signal Analysis Branches.

The *Administrative Office* is the Bureau's management office, handling personnel, budget, space, printing, anticipated work load, and similar areas. The *Program Development and Evaluation Staff* works closely with the Bureau Chief on short- and long-range policy development, legislation and rule-making.
(See Figure 9.)

Public Service Division

This division coordinates the Bureau's public information program, antenna survey matters, and licenses radio operators. It prepares bulletins, examinations, and issues licenses for radio operators. It reviews applications from other bureaus relating to antenna structures, making sure the marking and lighting are in accordance with the rules. It maintains liaison with the Federal Aviation Administration (FAA) on airspace matters, including antenna construction.

The *Antenna Survey Branch* reviews applications in all FCC-regulated services for proper marking and lighting of antenna structures. It maintains files on all antenna structures and provides assistance to other offices on this subject.

The *Radio Operator Branch* administers the division's radio operator examinations and license issuing.

The *Public Contact Branch* handles complaints and inquiries, provides information to the public, and advises other offices in the Commission on public concerns.

Engineering Division

This office directs the Bureau's acquisition of land, structures, equipment, materials, and vehicles, and develops standards, procedures, and techniques for field measurements and observations. It also provides engineering, material, and service support to other bureau divisions.

The *Equipment, Construction, and Installation Branch* designs, constructs, and installs radio direction finders for FCC monitoring stations, develops specialized equipment not ordinarily obtainable from industry for field offices, establishes quality control procedures, and evaluates signal radiation detecting devices for field use.

The *Equipment and Standards Branch* establishes and implements technical measurement standards, acquires field equipment for monitoring, measuring, and direction finding, oversees contracts for construction of facilities, and conducts technical projects at field installations. It also prepares the bureau's engineering and equipment manuals.

Enforcement Division

This division deals with all phases of compliance by licensees with the technical rules and standards, including development of policies and rule-making; it does field inspections, investigations, monitoring, and measurements; checks possible interference from unauthorized and non-licensed facilities; issues violation notices; and coordinates activities of other bureaus, including international monitoring.

The *Management and Legal Assistance Branch* handles software, forms, and reports pertinent to enforcement, oversees military frequency coordination activities of field offices, and is responsible for the legal aspects of the enforcement program.

The *Investigations and Inspections Branch* establishes and manages procedures, guidelines and standards for the investigations and inspections parts of the enforcement program, is in charge of field assignments for new and experimental approaches, cooperates with other bureaus and government agencies in investigations and inspections, and manages several special enforcement tasks for the division.

The *Signal Analysis Branch* establishes and manages the technical procedures, guidelines, and standards for the monitoring enforcement program, directs unusual field approaches, coordinates with other bureaus and government agencies, and handles other special projects for the division.

Regional Offices

Each regional office is responsible for all FCC engineering activities in its region relating to station and wire facilities, including inspections, investigations, monitoring, operator examinations and licensing, identification and resolution of interference problems, information distribution, and assistance to public, private, and governmental offices in the region. It initiates corrective actions on interference problems, recommends policy and procedure changes in relation to enforcement, licensing, and spectrum monitoring, and establishes information systems in the region to assure proper implementation of FCC requirements.

4

▼
▼
▼
▼
▼

HISTORY

Although the first official regulatory agency, the Federal Radio Commission (FRC), was not established until 1927, the federal government became involved in international and domestic jurisdiction over communications systems almost a century earlier.

WIRE TELEGRAPH

Samuel F.B. Morse's invention of the electromagnetic telegraph in 1835 opened the door to distance communications, making it possible to transmit signals by wire from one point to another. In 1841, after years of unsuccessful attempts to get support for his invention, Morse received a grant from Congress to run a line from Washington to Baltimore—the government's first direct involvement in private-sector telecommunications. The telegraph's success in relaying the results of the Democratic National Convention in 1844 enabled Morse to raise enough private funds to extend the telegraph line to Philadelphia and New York. Telegraph companies were established in other parts of the country, as well. Western Union began in 1851 and built the first transcontinental telegraph line ten years later. The competitive Postal Telegraph Company began in 1881, and finally was merged with Western Union in 1943. Today, only Western Union offers a nationwide telegraph service.

While Morse's telegraph was crossing the land, he demonstrated in 1842 that signals could be sent also under water. It was Cyrus W. Field, however, who carried through the concept of an underwater cable linking Europe and the United States. In 1866, after a number of failed tries, a transatlantic link was connected in Newfoundland.

It didn't take long for the government to get involved. The first federal regulation of interstate electrical communication was with the passage of the Post Roads Act of 1866. It authorized the Postmaster General to fix rates annually for telegrams sent by the government. Citing the public interest, like it later did in the Radio Act of 1927 and the Communications Act of 1934, in 1887 the government authorized the Interstate Commerce Commission (ICC) to require telegraph companies to interconnect their lines for more extended public service.

WIRE TELEPHONE

Until 1875 all rapid long-distance communication depended on the telegraph. In that year, however, Alexander Graham Bell invented a new device, the telephone. With a patent issued in 1876, the first regular telephone line was constructed in 1877, from Boston to Somerville, Massachusetts. While telephone service was established between many key cities in the United States over the next 20 years, it wasn't until 1915 that the first transcontinental service by overhead wire was inaugurated.

The first Bell Telephone company had been founded in 1878; seven years later, in 1885, the American Telephone and Telegraph Company (AT&T) was incorporated. The first underground telephone cable linked New York and Newark, New Jersey in 1902, and it took 40 years more, until 1942, for the first cross-country underground telephone line to be in operation. In ensuing years underwater telephone lines and coaxial telephone cables (carrying multiple telephone, telegraph, or television signals) were developed.

RADIO TELEGRAPH

While the telegraph and telephone were in development, another equally far-reaching discovery was being made. In the 1860s a Scottish physicist, James Clerk Maxwell, predicted the existence of radio waves—that is, waves on which communication signals could be carried, similar to the signals that could be carried over the telegraph wire. In 1886 a German physicist, Heinrich Rudolph Hertz, projected rapid variations of electric current into space in the form of radio waves similar to those of light and heat. A patent for a wireless system was obtained in the United States in 1872. Little further progress was made in using this new invention to actually send communication signals until an Italian inventor, Guglielmo Marconi, sent and received a radio signal in 1895. In 1899 he sent the first wireless signal across the English channel and in 1901 the letter *S* was sent by wireless from England to Newfoundland. A year later Marconi sent the first transatlantic wireless message in the other direction, eastward. The signals were basically telegraph signals, without using wires. While early experiments were under way using wireless radiotelephone—that is, voice signals—wireless telegraph signals aroused world interest.

The most dramatic use of wireless communication was between ships and ship-to-shore, providing instant information for rescue ships and avoiding many disasters. In 1899 the United States Navy tried out wireless communication and in 1901 replaced its visual signaling and homing pigeon systems with the new medium. A number of experimental operations developed, including use by private, educational, and commercial interests, the Department of Agriculture, the Weather Bureau, and the Army and Navy. Even at that early date signal interference began to be enough of a problem that President Theodore Roosevelt established a committee in 1904 to deal with that and other concerns. The committee's recommendations included a number of regulatory suggestions that were not adopted at that time, but established the principles that were to prompt regulation over the next two decades. During   that decade radiotelegraph service grew, connecting countries, covering world events, and reporting weather conditions. In 1909 arctic explorer Robert E. Peary used the radiotelegraph to tell the world "I found the Pole." In 1910 Marconi started regular American-European radiotelegraph service, and in 1912 San Francisco was linked with Hawaii.

In World War I governments experimented with the radiotelegraph to monitor battlefield strategy and to direct the movements of troops and supplies. But its full potential was not yet understood or exploited. The first pictures were sent by radio-telegraph in 1923, and transatlantic photo relay was inaugurated in 1924.

By 1910, however, wire communication carriers had grown and the government was concerned about their monopoly and other business practices, including their service to the public. In that year the Mann Elkins Act was passed, which gave the government regulatory power over accounting practices of radiotelegraph carriers. The ICC was authorized to establish uniform systems of accounts for telegraph and telephone carriers, to make valuation studies of certain companies, and to receive information from the companies of extensions and improvements in their operations. The Act also designated the ICC as a regulatory agency over radiotelegraph carriers, and extended the provisions of the Interstate Commerce Act of 1887 that gave the ICC jurisdiction over interstate activities of wireless telegraph.

At the beginning of the 20th century the radiotelegraph's use in protecting life and property at sea had become so widespread throughout the world that in 1903 an international conference was held in Berlin to discuss common distress call signs for ships and to promote wireless communication between ship and shore—which was not yet in practice—as well as between ships. In 1906 "SOS" was adopted as an international radiotelegraph distress call, and is still in use today. In 1927, when radiotelephony was in common use, "Mayday" was adopted as the voice equivalent.

In 1910 the United States Congress enacted the first legislation dealing with marine radio, the Wireless Ship Act, which required installation of wireless apparatus and operators on large sea-going passenger vehicles. This Act is considered by many to be the true forerunner of the 1927 and 1934 Acts. It came about because the United States had not yet become a signatory to the Berlin Conferences (a second one was held in London in 1906), and the international community withdrew an invitation to the United States to attend a third conference in 1912. Congress quickly passed the Act, which adopted the international agreement. The Secretary of Commerce, who at that time administered the domestic maritime navigation laws, was given responsibility for enforcing the Wireless Ship Act. In 1912 the Act was amended to expand its requirements to more ships and strengthen radio services and communications aboard vessels.

That same year, 1912, the Berlin International Radiotelegraphic Convention met in London and enacted international regulations to further wireless conformity and compatibility. To comply with these regulations, the United States Congress enacted the Radio Act of 1912. This is generally considered to be the first law in this country regulating radio communications. The specific requirements foreshadowed the detailed rules and regulations of later regulatory Acts and Commissions. The Radio Act of 1912 regulated the character of emissions and the transmissions of distress calls, set aside certain frequencies for government use, and established licensing of wireless stations and operators, placing the implementation under the Secretary of Commerce and Labor. The Act stated, in part:

> ... that a person, company or corporation within the jurisdiction of the United
> States shall not use or operate any apparatus for radio communication ... except
> under and in accordance with a license ... granted by the Secretary of Com-
> merce and Labor ... That every such license shall be in such form as the Secre-
> tary ... shall determine and shall contain restrictions ... that every such license

shall be issued only to citizens of the United States . . . shall specify the ownership and location of the station . . . to enable its range to be estimated . . .shall
state the purpose of the station . . . shall state the wavelength . . . authorized for
use by the station for the prevention of interference and the hours for which the
station is licensed to work . . . .

RADIO TELEPHONE

No one knows for certain when the first human voice was communicated over
the airwaves. Some claim the first such transmission was in 1892 when inventor
Nathan B. Stubblefield spoke the words "Hello Rainey" to an assistant a distance
away in an experiment near the town of Murray, Kentucky. Others insist that the true
predecessor of modern radio didn't take place until 1906 when inventor Reginald A.
Fessenden broadcast a program of talk and music from Brant Rock, Massachusetts,
that was picked up by ships as far as several hundred miles away. Speech was first
transmitted across the continent, from New York to San Francisco, and across the
ocean, from Arlington, Virginia, to the Eiffel Tower in Paris, in 1915. Strangely, the
United Armed Forces did not take much advantage of this wireless speech in World
War I, trying it out primarily for airplane-to-ground purposes.

In 1922 the first ship-to-shore two-way conversation took place, and in 1929
public radiotelephone service on ships was established. Commercial circuits between
the United States and Europe began in 1927, and in 1935 the first telephone call was
made around the world, using both wire and wireless.

Radiotelegraphy, radiotelephony, and the telephone were the communications
media that principally occupied the government and most of the public from the pre-
to the post-World War I period. From August 1, 1918 to July 31, 1919, the government took control over telephone and telegraph communications as a war measure.
In 1920 Congress authorized the Secretary of the Navy to use government-operated
stations for the transmission of press and private commercial messages between ship
and shore at reasonable rates subject to review by the ICC. The Transportation Act
of 1920 authorized the ICC to prescribe depreciation rates and charges of telephone
and telegraph companies. That same year the Interstate Commerce Act was amended
to permit consolidation of telephone companies when approved by the ICC. In 1921
an Executive Order authorized the Department of State to receive all applications to
land or operate ocean cables, and to advise the President of the granting or revoca-
tion of such licenses.

While radiotelegraphy and radiotelephony were predominant, something
new was emerging from the advancing technology: broadcast radio. Although
Stubblefield and Fessenden are credited with some of the key initial demonstrations
of radio, Lee De Forest is generally considered to be the father of American radio.
The electron tube is the basis for AM radio. In 1883 Thomas Alva Edison had observed the emission of electrons from a heated surface, such as a tube's cathode, but
the electron itself had not yet been discovered; that is credited to British researcher
Sir J.J. Thomson through a series of experiments in the 1890s. In 1904 English engineer Sir John A. Fleming discovered an application for the simplest electron tube, the
diode. But the diode couldn't amplify the electronic signals. Lee De Forest's inven-

tion of the triode, which he called the audion, a tube that could amplify the signal, opened the way for the development of radio as we know it today. De Forest's patents and a number of others by key inventors and researchers of the time were eventually put into a patent pool which, through litigation, politicking, and, according to some historians, corruption and bribery, eventually proved the basis for the formation in 1920 of the Radio Corporation of America, which became the single most important force in the development of broadcasting. The government's interest in facilitating the growth of radio resulted in its turning its back on what were many perceived violations of anti-trust laws by the newly formed media corporations.

BROADCASTING

The Radio Act of 1912 did not anticipate or provide for broadcasting, which presented no problem prior to World War I; early broadcasting was experimental, principally conducted by professors and students in physics and engineering departments of colleges and universities. Early informal broadcasts served practical needs as well as theoretical classroom purposes, such as informing farmers through regular and irregular scheduled radio reports of weather conditions, crop information, produce prices, Department of Agriculture findings and advisories, and other things that the isolated farmers otherwise sometimes would have to wait for days to learn.

After World War I a number of scheduled radio broadcasts by some educational stations, and the establishment of experimental radio stations by businesses, factories, department stores, hotels, and other commercial interests made it clear that radio was here to stay. In 1919, after the government preemption of stations under war powers was ended, so many experimental stations went on the air that the Department of Commerce, which had previously been given jurisdiction over the generic concept of radio in the Radio Act of 1912, issued a formal classification of these operations as "limited commercial stations." While there is controversy as to which station actually was the first one on the air, the claim of KDKA, Pittsburgh, is accepted by most historians. On November 2, 1920, KDKA broadcast the national election returns, what is conceded to be the first formal radio program of a regular schedule.

The Secretary of Commerce was authorized to license stations. However, licensing was a pro-forma procedure because the Radio Act of 1912 gave the Department of Commerce no additional legal authority. The first regular broadcast license was issued to WBZ, Springfield, Massachusetts, on September 15, 1921; KDKA did not receive its license until later in the year, on November 7. Stations began to go on the air as they wished. The procedure was to go to Washington, file an application with the Department of Commerce in the morning, eat lunch, come back in the afternoon and receive the license, and go back home that same day.

In 1922 a wavelength of 360 meters (about 830 kc) was assigned for the transmission of "important news items, entertainment, lectures, sermons, and similar matter." Until then there was no advertising. Manufacturers of electronic equipment used in the construction of stations and receivers were anxious to sell their new products; but people wouldn't buy sets unless there were stations to listen to. So many of these manufacturers set up radio stations so people would have entertainment that would prompt them to buy sets, which in turn would prompt more stations to go on the air. Department stores set up stations in order to draw customers to the broadcast site. Hotels did the same thing. All broadcasting was live, and people for the first time were able to hear at home—other than on records—some of the bands and musical personalities they could hear before only by paying to go night clubs or vaudeville theaters. Some studios were very plain—literally a broom closet—and some later became quite ornate.

Chaos on the Air

The biggest spur to the development of radio came on August 28, 1922, when station WEAF, New York, broadcast the first commercial—for a new apartment complex in Queens. Secretary of Commerce Herbert Hoover and many others associated with government and the development of radio had characterized radio as an educational medium and had both argued and assumed that it would not be used for commercial advertising purposes. In 1924, at the Third Radio Conference (see below) Secretary Hoover said: "I believe that the quickest way to kill broadcasting would be to use it for direct advertising. The reader of the newspaper has an option whether he will read an ad or not, but if a speech by the President is to be used as the meat in a sandwich of two patent medicine advertisements there will be no radio left."

Increasing numbers of stations prompted Hoover to call a national meeting of parties interested in the growth of radio. The First National Radio Conference was held in Washington, D.C., in 1922, and resulted in further ostensible regulations: a new type of broadcast station with minimum power of 500 watts and maximum power of 1000 watts, with additional frequencies assigned.

Stations continued to proliferate, and without Commerce Department authority to assign specific frequencies or power in given geographical areas, a virtual Tower of Babel resulted, with stations even in the same city going on the same frequency and drowning out themselves and other stations. Pretty soon, in many areas of the country, there was chaos on the air, with fewer and fewer stations able to reach the public without interference. Broadcasters simply changed frequencies and increased power and hours of operation as they wished. Three more national radio conferences were called in successive years—1923, 1924, and 1925—to try to solve the problem. One action following the 1924 conference was the expansion of the frequencies allocated for radio broadcasting to the present AM band of 550-1,500 KHz, with power up to 5,000 watts. Still the interference continued and the 1925 conference asked the Secretary of Commerce to limit broadcast time and power. He was prevented from doing so by court cases that held that the Commerce Department had no legally designated authority to do so under the Radio Act of 1912. In 1926 President Calvin Coolidge asked Congress to remedy the chaos on the air. The result was the Dill-White Radio Act of 1927.

The Federal Radio Commission

The Radio Act created a five-member Federal Radio Commission with regulatory powers over radio, including the issuance of licenses, the allocation of frequency bands to various services, the assignment of specific frequencies to individual stations, and the designation of station power. In addition, the Secretary of Commerce was authorized to inspect radio stations, examine and license radio operators, and assign radio call signs. The Radio Act also contained provisions that established the principle of regulatory action in the public interest in regard to some program matters that exploited or deceived the listener, and the FRC acted against religious charlatans whose principle purpose was milking the public for donations, patent medicine hucksters, fortune-tellers, and others. Although Congress did not define what it meant by the "public interest, convenience, and/or necessity"—and has not done so to this day—the statement established the base for later regulation that went far beyond that of technical supervision.

The Radio Act of 1927 and the establishment of the Federal Radio Commission solved the chaos on the air by reorganizing radio assignments and licensing, causing some 150 of the 732 stations then on the air to surrender their licenses. In 1928 the FRC issued an order that made full use of the AM broadcast band and established classes of stations in geographical zones. The movement of the FRC into certain areas of programming prompted the National Association of Broadcasters (NAB), which had been established in 1923 primarily to deal with the royalty demands of the American Society of Composers, Authors, and Publishers (ASCAP), to try to pre-empt stronger government regulation by self-regulation, and in 1929 it established a Code of Ethics, which continued in various forms until the 1980s.

Programming was becoming more sophisticated, and stars from other fields— vaudeville, burlesque, the theatre, nightclubs—were beginning to appear on radio. While as early as 1922 the concept of networks or, as then called, chain broadcasting, was tried out by connecting two stations, then several stations, in 1926 the National Broadcasting Company, a subsidiary of the Radio Corporation of America, started the first regular network with 24 stations. Its first coast-to-coast hookup was in 1927, the same year the other still-active broadcast giant, the Columbia Broadcasting System, was organized. With national instant exposure, advertisers began to flock to radio, and with money and the same exposure, so did more and more big-name entertainers.

While the Radio Act established a regulatory body for radio, a new medium was already in operation, television.

Television and Other New Technologies

Television had been on the horizon for many years, following a number of electronic discoveries in the late 19th and early 20th centuries. In 1884 German inventor Paul Nipkow patented a scanning disc for transmitting pictures by wireless. American Charles F. Jenkins began researching television potentials in 1890. English physicist E.E. Fournier d'Albe conducted experiments in the early 1900s. By 1915 Marconi was predicting the "visible telephone." During the next decade television developed quickly with iconoscope tube patents applied for in 1923. Leading experimenters in the United States were Vladmir Zworykin, Herbert E. Ives, Charles Francis Jenkins, E.F.W. Alexanderson and Philo T. Farnsworth. The latter is generally considered to be the father of American television, developing the electronic, as differentiated from the mechanical, TV system. The first demonstration of television took place in London in 1926 by English inventor John L. Baird. In 1927 the first test of television in America carried an experimental hookup by wire between New

York and Washington featuring Secretary of Commerce Hoover. Bell Telephone Laboratories was responsible for that experiment and by the following year was televising outdoor programs. The General Electric Company in Schenectady, N.Y., was operating what could be described as an experimental television station, WGY, and televised the first TV drama in 1928. By 1930 large-screen television was being demonstrated by RCA at a theater in New York.

The Radio Act of 1927 had established the FRC for only a 1-year period as a regulatory body, but Congress extended its authority on a continuing basis. Yet, even as radio was finally being regulated, new and expanded electronic communication developments were making the FRC quickly outdated. Television and still another new medium were about to enter the American communications mainstream. Edwin H. Armstrong, who had been a young inventor on the cutting edge of early AM radio, was developing a new, static-free radio system he called FM. Further, the Radio Act of 1927 did not give the FRC jurisdiction over telegraph and telephone carriers; telegraph supervision was divided among the Post Office Department, the Interstate Commerce Commission, and the Department of State; the ICC had some jurisdiction over telephone, with the FRC having authority over the broadcasting part. This divided and overlapping situation resulted in much confusion, duplication, and inefficiency.

THE FEDERAL COMMUNICATIONS COMMISSION

In 1933 President Franklin D. Roosevelt requested the Secretary of Commerce (no longer Hoover, who had been elected President in 1928, and was defeated by Roosevelt after serving one term) to appoint and convene an interagency committee to study the problem of government regulation of electronic communications. The interdepartmental committee recommended that congress establish a single agency to regulate all interstate and foreign communications by wire and radio, including broadcasting, telegraph, and telephone, with provisions for inclusion of newly developing technologies and media that might fall under these categories. Congress enacted the Communications Act of 1934, which created the Federal Communications Commission to be responsible for this unified regulation. The FCC began operations on July 11, 1934, as an independent agency comprised of seven Commissioners, appointed by the President with the advice and consent of the Senate; in 1983 the number of Commissioners was reduced to five. Chapter I of this book describes the duties and organization of the FCC under the Communications Act of 1934.

Since 1934 the regulatory powers and practices of the FCC have changed principally only in degree, its basic mandate still governing its activities. The expansion of existing communication technologies and the development of new ones, such as satellite, cable, and others previously discussed, resulted in new rules and regulations and, in some instances, amendments to the Communications Act. Depending on the party in power, the country's political climate, and the philosophical orientation of individual Commissioners, the FCC has regulated strongly in the public interest or deregulated in the industry interest. For example, the FCC's issuance of the so-called *Blue Book*—officially called *Public Service Responsibilities of Broadcast Licensees*—in 1946 outlined station program responsibilities in the public interest and established the FCC's authority to see that the stations lived up to those responsibilities. This occurred in a national euphoria of democracy following America's victory for the "people" over the "fascists" in World War II. FCC rulings requiring stations to demonstrate their public interest commitments were epitomized by the "vast wasteland" concerns of President John F. Kennedy's 1961 appointee as FCC chair, Newton F. Minow. The 1960s and part of the 1970s were an era of strong regulation. This coincided with a national sense of "people power" that included the Civil Rights revolution, the beginning of the women's movement, and public action that forced an end to American involvement in the war in Southeast Asia. Conversely, the 1980s saw deregulation of the communication industry in accordance with the marketplace theories of President Ronald R. Reagan, reflecting the strong support he received from the public in the *me generation* era that stressed personal material aggrandizement as differentiated from humanistic concerns. Many of the key elements in the FCC's regulatory and deregulatory modes are noted throughout this book in discussions of specific services and issues under FCC jurisdiction.

Early in its existence the FCC turned its attention to two long-standing problems, program content and network monopoly. It was concerned with false and proliferated advertising and with obscenity and profanity. In 1936 it began an investigation into AT&T's rate structure; in 1938 it started an inquiry into network practices, and in 1941 it made a far-reaching decision in adopting rules forbidding one organization from operating two networks—a forerunner of the later multiple-ownership rules, discussed earlier. The National Broadcasting Company was forced to divest itself of one of its two networks; NBC sold its Blue network to what eventually became the American Broadcasting Company, and retained its Red network, which continued as NBC. Years later, after television changed the structure and programming of radio, NBC, ABC, CBS, and the Mutual Broadcasting System, among others, were granted permission to operate multiple network operations in terms of distribution of different radio formats.

Although forbidden by the Communications Act from censoring, the FCC could establish general guidelines for programming. In 1939 it issued such guidelines—not as rules that could be enforced with sanctions, but strong enough so stations knew the FCC could give them a hard time on other matters if they were in violation—regarding advertising, liquor, religious and racial bigotry, obscenity, excessive music programming to the detriment of other types, and an area that was to evolve into the Fairness Doctrine, the presentation of various viewpoints on controversial issues.

By the mid-1930s there were over a dozen experimental TV stations operating in the United States, and in 1939 a public display of television was inaugurated by President Roosevelt at the New York World's Fair. On April 30, 1941, the FCC issued the first commercial TV authorizations, adopting a technical standard recommended by the National Television Standards Committee that has given the U.S. the lowest quality of the three systems in the world. It remains today a principal reason for the pressure to adopt and implement a high-definition television standard that would considerably improve reception quality. When the United States entered World War II, new and expanded station authorizations were suspended by the FCC and most of the dozen or so TV stations that were operating went off the air. By

1945, with the end of the war near, some had gone back on and a backlog of over 150 TV station applications was pending. At the same time, FM radio, which as the war started had some 400,000 receivers in use throughout the United States, had 600 applications pending.

The FCC's authorization of FM radio in 1940 barely got that new medium on the air when the ban on new construction during World War II left FM with only about 40 stations in operation by the war's end. In 1945 the FCC moved FM to its present higher-frequency band, 88–108 MHz, and increased the number of usable channels to 100–80 for commercial use, the remainder for nonprofit educational licensees. Edwin Armstrong wanted a better frequency range for FM—the same one the FCC was considering for TV audio.

By 1948 the postwar communication explosion prompted the FCC to put a freeze on all new TV applications. In 1952, in its Sixth Report and Order, the FCC, among other things, established UHF channels; reserved TV channels for noncommercial educational stations; specified mileage separation distances for television stations in order to reduce interference potentials; made city-by-city TV assignments; and assigned the more desirable audio frequencies for TV sound transmission. (This left FM where it was, with frequencies its founder, Edwin Armstrong, believed were inadequate for the effective growth and full service of FM. This action, coupled with exhausting royalty battles with RCA, prompted Armstrong's suicide two years later— before he could see the subsequent growth of FM that led it to surpass AM and to reach its present poisition.) The following year, in 1953, it adopted a color television system. The 1950s were a time of technological, station, and programming growth. They were also a time when the broadcast industry reached its lowest level of ethics, courage, and honesty.

The Blacklist   The "cold war" that followed the hot war of 1939-1945 was abetted by the media's exacerbating the confrontational atmosphere between the American and Soviet governments and people. The "red scare" was fueled by McCarthyism, the demagoguery and fear associated with Senator Joseph McCarthy. Broadcasting capitulated to an organization called AWARE, whose report, *Red Channels*, listed performers, writers, directors, and others who were allegedly "left-wingers." All the networks agreed to blacklist these people and pay the blacklisters fees to check the names of prospective performers on programs. Hundreds of careers were ruined on the basis of accusation alone. The blacklist ostensibly ended when a blacklisted CBS star, John Henry Faulk, won a multi-million dollar lawsuit against AWARE in 1962, although a "graylist" continued for some years afterward.

Some critics and historians believe the FCC and the federal government should have acted then, as they did at a later date with equal opportunity laws and rules to prevent stations and networks from discriminating on the basis of race or gender, to prevent broadcasters from discriminating on the basis of alleged political beliefs.

Other scandals   Malfeasance in office—specifically taking gifts from the broadcasters they were supposed to regulate—resulted in the resignation of two FCC Commissioners in the late 1950s and early 1960s. The broadcast industry itself was investigated by the FCC for rigging quiz shows and for disc jockeys accepting payments (payola) to promote certain records. The FCC established new rules to prevent such occurrences in the future.

Regulatory Atmosphere

The 1960s saw the beginnings of conscientious FCC implementation of the "public interest, convenience, and necessity" provision of the Communications Act, lasting well into the 1970s, with citizen groups encouraged to become involved in FCC licensing and renewal processes, and with stronger rules and regulations requiring stations to serve their community's needs. An Ascertainment of Community Needs rule required stations to determine the ten key issues in their communities and each year report to the FCC on how they programmed to disseminate information on those issues. The Prime Time Access Rule (PTAR) limited stations to 3 hours of network programs (with the exception of news) during the 7 to 11P.M. (EST) period, opening the door for independent program producers. Minimum time percentages per week were set for news and public affairs programs. Syndex (the syndicated exclusivity rule) limited network financial control to 50% of the programs they carried. The anti-monopoly rules, discussed earlier, were developed. Equal opportunity and affirmative action requirements were established, as were minority preference guidelines in obtaining licenses. The Fairness Doctrine (discussed in Chapter 3), which developed in a number of court cases beginning with the Mayflower Decision in 1941, the FCC editorializing report in 1949, the Supreme Court's Red Lion Broadcasting decision in 1969, and subsequent cases, gave the FCC the authority to expand the First Amendment to many more voices and opinions in the country— although the broadcast industry claimed the opposite, that the Fairness Doctrine hindered its First Amendment rights.

In 1975 the FCC revoked the licenses of the Alabama Educational Television network because of its racist practices in programming and employment. The FCC began its first study of cable in 1959, issued interim regulatory statements over the years, and in 1972 issued definitive rules clarifying its authority to regulate cable. Pay-TV came on the horizon and initial experiments were authorized by the FCC. A key broadcasting development occurred in 1967, when the Communications Act was amended to include the Public Broadcasting Act, which established the Corporation for Public Broadcasting (CPB) and authorized tax monies for CPB to develop national TV and radio systems (which it did with PBS and NPR, as discussed in Chapter 3). The FCC was given no regulatory authority over CPB or its offshoots.

From the beginning of television, children's television has been addressed by the FCC. While the FCC considered developing rules regarding children's television many times, it never did so. The closest it came was in 1974, when it issued suggested guidelines for children's shows; but because only violations of rules could result in sanctions, compliance with the FCC's suggestions was totally voluntary. In 1990 Congress limited commercial time on children's TV programs and required the FCC to consider the quality of such programming at renewal time.

The developing technologies, including cable, satellite, cellular radio, fiber optics, microwave services, low power TV, high-definition TV, and other emerging telecommunication techniques were major concerns of the FCC from the late 1970s into the 1990s. Although, as noted earlier, the Cable Communications Act of 1984 removed most of the FCC's jurisdiction over cable, some remained. Most important was the FCC's concern with the "must-carry" rule, which had required cable systems to carry local television channels; when this was declared unconstitutional the FCC tried to reenact it, to no avail.

Deregulation   The major philosophical change in the FCC was the move from regulation to deregulation. In reaction to what many considered to be an overload of government requirements and paperwork, the Nixon-Ford administrations' FCC chair, Richard Wylie, began what was termed re-regulation during his tenure from 1972 into 1977. This was accelerated under President Carter's FCC chair, Charles Ferris, who in his tenure from 1977 into 1981 moved the FCC further away from the heyday of public-interest regulation toward what many critics regard as the pro-industry, anti-consumer deregulation of the following decade. President Reagan's marketplace philosophy of extreme, almost total deregulation took place under his FCC chair, Mark Fowler, from 1981 into 1987. Virtually all of the pro-consumer regulations of the 1960s and 1970s, noted above, were eliminated. Fowler's successors have maintained the deregulatory philosophy, although President Bush's FCC chair, Alfred Sikes, who began his term in 1989, seems to have slowed down the pace toward nonregulation.

Throughout the history of the FCC it has been accused of being a reluctant regulator. It is an axiom, though not necessarily a truism, in Washington that every regulatory agency is controlled by the industry it is supposed to regulate. Many people have the conception that the FCC is dominated by the broadcast industry. Cable advocates who have closely observed the FCC's actions in cable-broadcasting controversies frequently voice this opinion. Except for the Jack Kennedy "legacy" period from the early 1960s to the mid-1970s, consumer and citizen groups frequently make the same accusation. Some critics state that as long as FCC commissioners are appointed on the basis of political reward rather than experience and competence in communication fields, industry influence will continue. Others insist that the practice of former commissioners, with few exceptions, in obtaining jobs or consultancies with the communications industry after they leave the FCC, makes the bias obvious. Still others disagree and say the FCC is as honest and unbiased as any federal bureaucracy agency can be.

At various times since 1934 attempts have been made to rewrite the Communications Act of 1934, most notably in the late 1960s, when President Lyndon Johnson appointed a committee to develop a plan and rationale for such a rewrite, and in the late 1970s when the House Communications Subcommittee, chaired by Representative Lionel van Deerlin, prepared a 1977 report on regulatory status and needed change. New Communication Act bills introduced by Van Deerlin in 1978 and again in 1979 were not approved by Congress. At this writing it does not appear that a major revision of the Communications Act is likely in the near future.

5

▼
▼
▼
▼
▼

Process and Procedure

HOW FCC RULES ARE MADE

Sources

New FCC rules and regulations or changes in the existing ones may be initiated from many sources, including individual citizens. Any person, group, or organization may petition the FCC to enact or change a specific rule. The petition (which, at the present time, must be formally submitted to the FCC Secretary as an original with 14 copies) details the reasons for the recommended rule and states the specific language desired for the rule. Over the years there have been a number of occasions when a petition for rule-making initiated by an individual resulted in a new rule.

Rules can also be mandated by legislation. Any time Congress amends the Communications Act of 1934, the language of the amendment may include specific regulations to be implemented by the FCC. The Executive Branch frequently is responsible for new rules. Although the White House ostensibly stays at arm's length from independent agencies such as the FCC, special committees, other departments, agencies and offices under the Executive Branch, and individuals serving directly under the President may informally let the FCC know of their interest in seeing a new or changed rule. Such suggestions are always given serious consideration by the Commission and frequently are acted on. The third branch of government, the Judiciary, is also responsible for initiating new rules. This comes about as a result of court decisions, usually in challenges to an existing FCC rule, that find it necessary to eliminate, clarify, or strengthen a rule. For example, in the 1980s the federal courts twice found that the FCC's cable "must-carry" rule was unconstitutional, requiring the FCC to eliminate or modify it. In 1990 the courts were reviewing the FCC's cross-ownership and obscenity rules which, as you read this, may already have been changed.

The FCC itself may initiate rule changes. Any bureau or office in the Commission may prepare a rule change, drawing up a draft of the rule-changing document, clearing it through the appropriate bureau chief, and then testing it out with the commissioners, a majority of whom may decide to put it on the official weekly agenda.

Sometimes such internal rule-making procedures are initiated from a commissioner's office. The Chairman of the FCC is the key figure, with the strongest influence in this procedure.

Evaluation and Process

When the FCC receives a petition for rule-making, it is sent to the bureau or office under whose jurisdiction the particular subject falls. If that office decides that the petition is meritorious—that is, is not frivolous and establishes a reasonable need for the proposed change—it can request that the FCC Dockets office assign it a *Rule Making (RM)* number. As noted above, rarely does an office decide on its own to take this step, but clears it first with appropriate higher offices, specifically those of the commissioners and FCC chairman. The same procedure holds true for changes initiated from sources other than the public, such as other arms of government or the FCC itself, as described above. Once the process is under way, with an RM number assigned, the petition or proposed rule making generated internally is listed along with all others on a weekly public notice. The public has 30 days in which to submit comments to the FCC. Following reception and analysis of the comments, the FCC develops it as an agenda item to be considered by the Commission. The office that prepares the item usually recommends that the Commission issue one of four determinations: a *Memorandum Opinion and Order (MO&O)* denying the petition or ending any further action on the item; a *Notice of Inquiry (NOI)*, in which the FCC issues a statement describing the problem and asking for public comments on how the problem should be solved; a *Notice of Proposed Rule Making (NPRM),* in which the Commission states exactly how it proposes to change the *Rules*; or a *Report and Order Adopting Change*, which at this stage of the process is limited to changes in the editorial content of rule, but not changing its substance in any material way. If an NOI or NPRM is issued, it must be accompanied by a docket number that follows that particular item until it is finally resolved. An NOI must be followed by an NPRM or by an MO&O concluding the inquiry.

Following the issuance of an NOI or NPRM the public is given a specified period of time in which to submit *comments*, and then, after all the comments are in, another period of time in which to submit *reply comments*. Comments may come from the public at large; usually, of course, they come from parties especially interested in the particular rule change, principally from industry, professional associations, and concerned citizen/consumer groups. If the FCC doesn't receive enough comments to make an informed decision, it may issue another NOI or NPRM, again asking for comments and reply comments. Sometimes the Commission may hold an *oral argument* meeting, in which interested parties are invited to present their viewpoints in person. In all cases, whether at the initial agenda meeting adopting the item, or at oral arguments, representatives from the concerned FCC bureau or office present their viewpoints orally.

Following evaluation of comments and reply comments on an NOI, the Commission may issue an NPRM, continuing the matter, or an MO&O, concluding the inquiry. With an NPRM the next step is a Commission decision, in which it issues a

Report and Order (R&O) stating what the new or amended rule is, or in which it decides not to adopt or change the rule, thus terminating the proceeding, either in whole or in part. If in part, work on the item continues, and the FCC may issue a number of succeeding Reports and Orders, each one disposing by termination or adoption a part of the item. As noted in Chapter 4, one of the FCC's most significant proceedings began with the 1948 freeze on television and ended in 1952 only after a Sixth Report and Order on the item established new classes of television stations and a number of other new and critical regulations.

A further step in the process may occur. Following the issuance of an R&O, any member of the public may, within 30 days, file a *Petition for Reconsideration* (of the Commission's determination). The FCC reviews the petition and may issue one of two further determinations: an MO&O modifying its R&O in part or whole in accordance with the petition for reconsideration, or denying the petition.

LICENSING

How to Get a License

Few services regulated by the FCC do not require a license of some kind. As noted in Chapter 3 descriptions of the Private Radio and Common Carrier Bureaus, some operations may be undertaken without a license. However, even though a license may not be required, the equipment must conform to FCC standards and be operated in such a way as to not cause interference with other users in that frequency band, and the operation must conform to any applicable FCC rules. The outstanding example of a service not needing a license is Citizens Band Radio. In the early 1970s when these "walkie-talkies" became a national fad, licenses were required; but literally millions of applications were received, forcing the FCC at first to rent additional offices near Washington, D.C. in which to process and store these applications, and subsequently to simply eliminate the need for a license. Ostensibly, all citizen band operators are supposed to inform the FCC that they are operating such equipment, but without any possibility of realistic enforcement the rule has remained dormant.

Mass Media Application Forms and Fees   Any party wishing to construct a new station, make changes in an existing station, or renew or transfer a license must file the appropriate written application with the FCC on the prescribed form. Most applications require a fee (initiated in 1987 as a means of meeting the FCC's budget requirements). The following chart lists the title of the application form, the proper form number, and whether a fee is required.

Application Title	*Form #*	*Fee Required**
Authority to Construct or Make Changes in a Commercial Broadcast Station	301	

cont.

Application Title	Form #	Fee Required*
TV—new and major changes		*
TV—minor changes		*
Radio—new and major changes		*
AM		*
FM		*
Radio—minor changes AM and FM		*
New Commercial Broadcast Station License	302	
TV		*
Radio		
AM		*
FM		*
AM Directional Antenna		*
Renewal of License for Commercial and Noncommercial Educational AM, FM, and TV Broadcast Stations	303S	
Commercial		*
Noncommercial		
Extension of Broadcast Construction Permit or to Replace Expired Construction Permit	307	
Consent to Assignment of Broadcast Station Construction Permit or License— TV and Radio—long form	314	
Consent to Transfer of Control of Corporation Holding Broadcast Station Construction Permit or License—TV and radio—long form	315	*
Consent to Assignment of Broadcast Station Construction Permit or License or Transfer of Control of Corporation Holding Broadcast Station Construction Permit or License TV and Radio—short form	316	*
Permit to Deliver Programs to Foreign Broadcast Stations	308	
Authority to Construct or Make Changes in an International or Experimental Broadcast Station	309	
International or Experimental Broadcast Station License	310	
Renewal of an International or Experimental Broadcast Station License	311	
Authorization in the Auxiliary Broadcast Services	313	*
Renewal of Auxiliary Broadcast License	313	*
Authority to Construct or Make Changes in a Noncommercial Educational Broadcast Station	340	
Authorization to Construct New or Make Changes in Instructional TV Fixed and/or Response Stations, or to Assign or Transfer Such Stations	330	

cont.

Application Title	Form #	Fee Required*
Instructional Television Fixed Station License	330-L	
Renewal of Instructional TV Fixed Station and/or Response Station(s) and Low Power Relay Station(s) License	330-R	
Authority to Construct or Make Changes in a Low Power TV, TV Translator, or TV Booster Station	346	*
Low Power TV, TV Translator, or TV Booster Station License	347	*
Consent to Assignment of a TV or FM Translator Station Construction Permit or License	345	*
Renewal of a Low Power TV, TV Translator, or FM Translator Station License	348	*
FM Translator or FM Booster Station License	350	
Authority to Construct or Make Changes in an FM Translator or FM Booster Station	349	
Cable TV Relay Service (CARS) Construction Permit/Modification/License Assignment/ Transfer of Control/License Renewal	327	*

In addition, many required authorizations are made without a formal application form, but through a written request. For example, there are no specified application forms for Direct Broadcast Satellite (DBS). Authorization to construct or make a major modification in a DBS, launch authority, and a license to operate all require fees.

In instances where an application in any of the services do not conform to FCC rules, or petitions to deny the application have been filed, or there are competing applications, the FCC may designate the application for hearing before an Administrative Law Judge. A commercial TV, AM, FM, or DBS applicant who wishes to continue to participate in the process must pay a fee for the hearing.

Mass Media Application Process   The procedure in applying for a broadcast station (TV, AM, FM) is essentially the same for other facilities such as ITFS, MDS, and auxiliary services. Any qualified citizen, company, or group may apply for authority to construct a station. They must show the FCC that they are legally, technically, and financially qualified, as prescribed in the *Rules*, and that the operation of the proposed station would be in the public interest. Professional engineering and legal services are extremely helpful, sometimes vitally necessary, in preparing an acceptable application.

Frequency Selection   Depending on the service, selecting a frequency differs. For an AM channel the applicant must do a frequency search in the selected community and locate an unused AM frequency that will not cause interference to or munity and locate an unused AM frequency that will not cause interference to or receive interference from existing stations or stations for which applications are already on file. As noted earlier, class of station and regional area are factors in finding a channel.

An FM applicant must find an unused channel from the FCC's Table of Allotments that assigns specific FM frequencies to specific communities. If no vacant channel exists in the community of choice, the applicant may do an engineering study, and if a non-assigned frequency is found that will not cause or receive interference, the applicant may petition the FCC to add that channel to the Table of Allotments; if the FCC does so, the new channel becomes available for applications during a period specified by the FCC. Once a channel is identified within the reserved band and the engineering study is completed, an application may be filed. Station selection is governed by various factors, including the six classes of stations and the zone restrictions on power and antenna heights. No special application is required for stereophonic broadcasting, either for FM or AM.

Television channel selection is like that for FM: the applicant must find an unused channel in the FCC's Table of Allotments. If no channel is available in the city of choice, a petition for a properly engineered *drop-in* channel may be filed, as with FM. TV applicants have a secondary choice: TV translators and Low Power TV. The latter may originate programming. Where interference results from full-power operation on an available or drop-in channel, the applicant may request an LPTV facility by showing that it will not result in interference.

Filing Procedure   Commercial applicants file on form 301, noncommercial educational applicants on form 340. The applications require information that proves (1) the applicant is legally qualified to be the licensee and operator of a station, as prescribed by the rules; (2) the applicant has enough guaranteed funds to build the station according to the specified costs and to begin operation; (3) the applicant's engineering specifications meet all the technical requirements of separation, interference, and type-accepted equipment; and (4) the projected programming, as outlined by the applicant, including a sample week's schedule, will serve the needs of the community. Under deregulation the latter factor has had less and less significance. Applications must be submitted in triplicate, with the appropriate filing fees attached.

All applicants must give local notice concerning their applications in newspapers of general circulation in the community in which the station will be licensed. Copies of the application must be maintained in a place available to the public, and the public must be afforded an opportunity to comment to the FCC on the application. Renewal applications follow the same procedure, except that the public notice is read on the station up for renewal.

Once the application is formally accepted for filing by the FCC it is given a number and is placed on a *cut-off* list. The cut-off list is released as a public notice by the FCC, opening up a 30 day period for the filing of any competing applications or petitions to deny. (Commercial FM applications follow a slightly different procedure, with a 45-day "window" period, a 30-day amending period, and then another

30-day period for opposition filings.) Depending on the backlog, it may take the given office processing the application anywhere from a few to many months to finish work on the application. If the application is complete and found to have no defects, petitions to deny, or competing applications, the application will be granted and construction may begin. If the application has problems or the applicant is not found to be qualified for licensee status, the application is returned—sometimes rejected completely, sometimes with an opportunity for the applicant to correct the defects and refile it. A Public Notice concerning the disposition of applications is always issued, and Petitions for Reconsideration of the FCC action—whether the application is approved or disapproved—may be filed within 30 days. As previously noted, there may be a hearing before an Administrative Law Judge (ALJ). A further step is for the Commission itself or its Review Board to hear oral arguments on the matter and confirm, modify, or reverse the ALJ's decision. An applicant has further recourse by appealing the Commission's final decision to the District of Columbia Appellate Court.

The following are examples of FCC application and reporting forms. Because of the lengths of some of the forms, some condensations have been made:

- Form 301, Application for Construction Permit for Commercial Broadcast Station (Figure 10). Included are Section 1, General Information; Section II, Legal Qualifications, pages 1 and 2 of 4 pages; Section III, FinancialQualifications; Section IV-A, Program Service Statement; Section IV-B, Integration Statement; Section V-A, AM Broadcast Engineering Data, pages 1 and 2 of 6 pages; Section V-B, FM Broadcast Engineering Data, pages 1 and 2 of 5 pages; Section V-C, TV Broadcast Engineering Data, pages 1 and 2 of 5 pages; Section VI, EEO Program, and Section VII, Certifications, are omitted.
- Form 574, Application for Private Land Mobile and General Mobile Radio Services (Figure 11). This form serves as the application for a number of types of stations, including mobile relay, private carrier, mobile, operational fixed, fixed relay, radiolocation land, and radiolocation weather radar, among others. Form 574, reproduced here in its entirety, is accompanied by a detailed instruction booklet that includes a listing of the codes and names for the various services applied for; that listing is included here.
- Broadcast Equal Employment Opportunity Program Report (Figure 12). Included are pages 1 and 4 of 5 pages.
- Ownership Report (Figure 13). Included are pages 1 and 3 of 3 pages.

Federal Communications Commission
Washington, D. C. 20554

FCC 301

Approved by OMB
3060-0027
Expires 2/28/92
See Page 25 for information
regarding public burden estimate

APPLICATION FOR CONSTRUCTION PERMIT FOR COMMERCIAL BROADCAST STATION

For COMMISSION Fee Use Only		For APPLICANT Fee Use Only
	FEE NO:	Is a fee submitted with this application? ☐ Yes ☐ No
	FEE TYPE	If fee exempt (see 47 C.F.R. Section 1.1112), indicate reason therefor (check one box): ☐ Noncommercial educational licensee ☐ Governmental entity
	FEE AMT:	FOR COMMISSION USE ONLY
	ID SEQ:	FILE NO.

Section I — GENERAL INFORMATION

1. Name of Applicant

Send notices and communications to the following person at the address below:

Name

Street Address or P.O. Box			Street Address or P.O. Box		
City	State	ZIP Code	City	State	ZIP Code
Telephone No. *(Include Area Code)*			Telephone No. *(Include Area Code)*		

2. This application is for: ☐ AM   ☐ FM   ☐ TV

(a) Channel No. or Frequency	(b) Principal Community	City	State

(c) Check one of the following boxes:

☐ Application for NEW station

☐ MAJOR change in licensed facilities; call sign: ....

☐ MINOR change in licensed facilities; call sign: ....

☐ MAJOR modification of construction permit; call sign: ....

File No. of construction permit: ....

☐ MINOR modification of construction permit; call sign: ....

File No. of construction permit: ....

☐ AMENDMENT to pending application; Application file number: ....

NOTE: It is not necessary to use this form to amend a previously filed application. Should you do so, however, please submit only Section I and those other portions of the form that contain the amended information.

3. Is this application mutually exclusive with a renewal application?   ☐ Yes ☐ No

If Yes, state:

Call letters	Community of License	
	City	State

▶ *Figure 10   Form 301. Application for construction permit for commercial broadcast stations.*

Section II – LEGAL QUALIFICATIONS

Name of Applicant

1. Applicant is: *(check one box below)*

☐ Individual   ☐ General partnership   ☐ For-profit corporation

☐ Other   ☐ Limited partnership   ☐ Not-for-profit corporation

2. If the applicant is an unincorporated association or a legal entity other than an individual, partnership, or corporation, describe in an Exhibit the nature of the application.

☐ Exhibit No.

NOTE: The terms "applicant," "parties to this application," and "non-party equity owners in the applicant" are defined in the instructions for Section II of this form. Complete information as to each "party to this application" and each "non-party equity owner in the applicant" is required. If the applicant considers that to furnish complete information would pose an unreasonable burden, it may request that the Commission waive the strict terms of this requirement with appropriate justification.

3. If the applicant is not an individual, provide the date and place of filing of the applicant's enabling charter (e.g., a limited partnership must identify its certificate of limited partnership and a corporation must identify its articles of incorporation by date and place of filing):

Date _____   Place _____

In the event there is no requirement that the enabling charter be filed with the state, the applicant shall include the enabling charter in the applicant's public inspection file. If, in the case of a partnership, the enabling charter does not include the partnership agreement itself, the applicant shall include a copy of the agreement in the applicant's public inspection file.

4. Are there any documents, instruments, contracts or understandings (written or oral), other than instruments identified in response to Question 3 above, relating to future ownership interests in the applicant, including but not limited to, insulated limited partnership shares, nonvoting stock interests, beneficial stock ownership interests, options, rights of first refusal, or debentures?

☐ Yes ☐ No

If Yes, submit as an Exhibit all such written documents, instruments, contracts, or understandings, and provide the particulars of any oral agreement.

☐ Exhibit No.

5. Complete, if applicable, the following certifications:

(a) Applicant certifies that no limited partner will be involved in any material respect in the management or operation of the proposed station.

☐ Yes ☐ No

If No, applicant must complete Question 6 below with respect to all limited partners actively involved in the media activities of the partnership.

(b) Does any investment company *(as defined in 15 U.S.C. Section 80 a-3)*, insurance company, or trust department of any bank have an aggregated holding of greater than 5% but less than 10% of the outstanding votes of the applicant?

☐ Yes ☐ No

If Yes, applicant certifies that the entity holding such interest exercises no influence or control over the applicant, directly or indirectly, and has no representatives among the officers and directors of the applicant.

☐ Yes ☐ No

▶ *Figure 10   (continued)*

Section II − LEGAL QUALIFICATIONS (Page 2)

6. List the applicant, parties to the application and non-party equity owners in the applicant. Use one column for each individual or entity. Attach additional pages if necessary.

(Read carefully - The numbered items below refer to line numbers in the following table.)

1. Name and residence of the applicant and, if applicable, its officers, directors, stockholders, or partners (if other than individual also show name, address and citizenship of natural person authorized to vote the stock). List the applicant first, officers next, then directors and, thereafter, remaining stockholders and partners.

2. Citizenship.

3. Office or directorship held.

4. Number of shares or nature of partnership interests.

5. Number of votes.

6. Percentage of votes.

7. Other existing attributable interests in any broadcast station, including the nature and size of such interests.

8. All other ownership interests of 5% or more (whether or not attributable), as well as any corporate officership or directorship, in broadcast, cable, or newspaper entities in the same market or with overlapping signals in the same broadcast service, as described in 47 C.F.R. Section 73.3555 and 76.501, including the nature and size of such interests and the positions held.

1.			
2.			
3.			
4.			
5.			
6.			
7.			
8.			

▶ *Figure 10* *(continued)*

SECTION III - FINANCIAL QUALIFICATIONS

NOTE: If this application is for a change in an operating facility do not fill out this section.

1. The applicant certifies that sufficient net liquid assets are on hand or that sufficient funds are available from committed sources to construct and operate the requested facilities for three months without revenue. ☐ Yes ☐ No

2. State the total funds you estimate are necessary to construct and operate the requested facility for three months without revenue. $ _____

3. Identify each source of funds, including the name, address, and telephone number of the source (and a contact person if the source is an entity), the relationship (if any) of the source to the applicant, and the amount of funds to be supplied by each source.

Source of Funds (Name and Address)	Telephone Number	Relationship	Amount

FCC 301 (Page 6)
June 1989

▶ *Figure 10* *(continued)*

Section IV-A - PROGRAM SERVICE STATEMENT

Attach as an Exhibit, a brief description, in narrative form, of the planned programming service relating to the issues of public concern facing the proposed service area.

> Exhibit No.

Section IV-B - INTEGRATION STATEMENT

Attach as an Exhibit the information required in 1. and 2. below.

> Exhibit No.

1. List each principal of the applicant who, in the event of a grant of the application on a comparative basis proposes to participate in the management of the proposed facility and, with respect to each such principal, state whether he or she will work full-time (minimum 40 hours per week) or part-time (minimum 20 hours per week) and briefly describe the proposed position and duties.

2. State with respect to each principal identified in response to Item 1. above, whether the applicant will claim qualitative credit for any of the following enhancement factors:

   (a) Minority Status
   (b) Past Local Residence
       If Yes, specify whether in the community of license or service area and the corresponding dates.
   (c) Female Status
   (d) Broadcast Experience
       If Yes, list each employer and position and corresponding dates.
   (e) Daytime Preference

▶ *Figure 10   (continued)*

Section V-A - AM BROADCAST ENGINEERING DATA	FOR COMMISSION USE ONLY
	File No. _____
	ASB Referral Date_____
	Referred by _____

Name of Applicant

---

1. Purpose of Application: *(check all appropriate boxes)*

☐ Construct new station

☐ Make changes in authorized/existing station   Call Sign _____

   ☐ Principal authorized/licensed community

   ☐ Frequency   ☐ Hours of operation

   ☐ Power   ☐ Transmitter location

   ☐ Main studio location

   ☐ Antenna system *(including increase in height by addition of FM or TV antenna)*

      ☐ New antenna construction

      ☐ Alteration of existing structure

         ☐ Increase height   ☐ Decrease height

         ☐ Non-DA to DA   ☐ DA to Non-DA

☐ Other *(Summarize briefly the nature of the changes proposed)*

---

2. Principal community to be served:

State	County	City or Town

3. Facilities requested:

Frequency: _____ kHz   Hours of Operations:

Power: Night: _____ kW   Day: _____ kW   Critical hours: _____ kW

4. Transmitter location:

State	County	City or Town

Exact antenna location *(street address)*. If outside city limits, give name of nearest town and distance *(in kilometers)*, and direction of antenna from town.

Geographical coordinates *(to nearest second)*. For directional antenna give coordinates of center of array. For single vertical radiator give tower location. Specify South Latitude or East Longitude where applicable; otherwise, North Latitude or West Longitude will be presumed.

Latitude ° ′ ″	Longitude - ° ′ ″

▶ *Figure 10*   *(continued)*

SECTION V–A – AM BROADCAST ENGINEERING DATA **(Page 2)**

5. Is the proposed site the same transmitter-antenna site of other stations authorized by the Commission or specified in another application pending before the Commission? ☐ Yes ☐ No

   If Yes, indicate call sign or application file number: _____

6. Antenna system *(including ground or counterpoise system)*

   Non–Directional   ☐ Day   ☐ Night   ☐ Critical Hours

   Estimated efficiency _____ mV/m per kW at one kilometer

   If antenna is either top loaded or sectionalized, describe fully in an Exhibit. | Exhibit No. |
   (Include apparent electrical height.)

   Directional   ☐ Day only (DA-D)   ☐ Night only (DA-N)

   ☐ Same constants and power day and night (DA-1)

   ☐ Different constants and/or power day and night (DA-2)

   ☐ Different constants and/or power day, critical hours and night (DA-3)

   Submit complete engineering data in accordance with 47 C.F.R. Section 73.150 for each Directional antenna pattern proposed.

   Type of feed circuits (excitation) ☐ Series Feed   ☐ Shunt Feed   ☐ Other (explain)

TOWERS *(In meters, rounded to nearest meter)*	1	2	3	4	5	6
Overall height of radiator above base insulator, or above base, if grounded						
Overall height above ground *(include obstruction lighting)*						
Overall height above mean sea level *(include obstruction lighting)*						

If additional towers, attach information exactly as it appears above.

7. Has the FAA been notified of the proposed construction?   ☐ Yes ☐ No

   If Yes, give date and office where notice was filed and attach as an Exhibit a copy of FAA determination, if available. | Exhibit No. |

   Date _____   Office where filed _____.

▶ *Figure 10   (continued)*

Section V-B — FM BROADCAST ENGINEERING DATA	FOR COMMISSION USE ONLY
	File No. _____
	ASB Referral Date_____
	Referred by _____

Name of Applicant

Call letters *(if issued)*

Is this application being filed in response to a window?   ☐ Yes ☐ No

If Yes, specify closing date: _____

Purpose of Application: *(check appropriate box(es))*

☐ Construct a new (main) facility

☐ Construct a new auxiliary facility

☐ Modify existing construction permit for main facility

☐ Modify existing construction permit for auxiliary facility

☐ Modify licensed main facility

☐ Modify licensed auxiliary facility

If purpose is to modify, indicate below the nature of change(s) and specify the file number(s) of the authorizations affected.

☐ Antenna supporting-structure height

☐ Effective radiated power

☐ Antenna height above average terrain

☐ Frequency

☐ Antenna location

☐ Class

☐ Main Studio location

☐ Other *(Summarize briefly)*

File Number(s) _____

1. Allocation:

Channel No.	Principal community to be served:			Class *(check only one box below)*
	City	County	State	☐ A ☐ B1 ☐ B ☐ C3
				☐ C2 ☐ C1 ☐ C

2. Exact location of antenna.

(a) Specify address, city, county and state. If no address, specify distance and bearing relative to the nearest town or landmark.

(b) Geographical coordinates (to nearest second). If mounted on element of an AM array, specify coordinates of center of array. Otherwise, specify tower location. Specify South Latitude or East Longitude where applicable; otherwise, North Latitude or West Longitude will be presumed.

Latitude ° ' "	Longitude ° ' "

3. Is the supporting structure the same as that of another station(s) or proposed in another pending application(s)?   ☐ Yes ☐ No

If Yes, give call letter(s) or file number(s) or both. _____

If proposal involves a change in height of an existing structure, specify existing height above ground level including antenna, all other appurtenances, and lighting, if any.

_____

▶ *Figure 10   (continued)*

SECTION V-B — FM BROADCAST ENGINEERING DATA (Page 2)

4. Does the application propose to correct previous site coordinates?   ☐ Yes ☐ No
   If Yes, list old coordinates.

Latitude ° ' "	Longitude ° ' "

5. Has the FAA been notified of the proposed construction?   ☐ Yes ☐ No
   If Yes, give date and office where notice was filed and attach as an Exhibit a copy of FAA
   determination, if available. | Exhibit No. |

   Date _____   Office where filed _____

6. List all landing areas within 8 km of antenna site. Specify distance and bearing from structure to nearest point of the
   nearest runway.

           Landing Area   Distance (km)   Bearing (degrees True)

   (a) _____   _____   _____

   (b) _____   _____   _____

7. (a) Elevation: *(to the nearest meter)*

      (1) of site above mean sea level;   _____ meters

      (2) of the top of supporting structure above ground (including antenna, all other   _____ meters
          appurtenances, and lighting, if any); and

      (3) of the top of supporting structure above mean sea level [(aX 1) + (aX2)]   _____ meters

   (b) Height of radiation center: *(to the nearest meter)*   H • Horizontal; V • Vertical

      (1) above ground   _____ meters (H)

                                                                                 _____ meters (V)

      (2) above mean sea level   [(aX 1) + (bX 1)]   _____ meters (H)

                                                                                 _____ meters (V)

      (3) above average terrain   _____ meters (H)

                                                                                 _____ meters (V)

8. Attach as an Exhibit sketch(es) of the supporting structure, labelling all elevations required | Exhibit No. |
   in Question 7 above, except Item 7(b)(3). If mounted on an AM directional-array element,
   specify heights and orientations of all array towers, as well as location of FM radiator.

9. Effective Radiated Power:
   (a) ERP in the horizontal plane   _____ kw (H*) _____ kw (V*)

   (b) Is beam tilt proposed?   ☐ Yes ☐ No

      If Yes, specify maximum ERP in the plane of the tilted beam, and attach as an Exhibit a | Exhibit No. |
      vertical elevational plot of radiated field.
                                          _____ kw (H*) _____ kw (V*)

      *Polarization

▶ *Figure 10   (continued)*

Section V-C – TV BROADCAST ENGINEERING DATA	FOR COMMISSION USE ONLY
	File No. _____
	ASB Referral Date _____
	Referred by _____

Name of Applicant	Call letters *(if issued)*

Purpose of Application *(check appropriate box)*:

☐ Construct a new (main) facility   ☐ Construct a new auxiliary facility

☐ Modify existing construction permit for main facility   ☐ Modify existing construction permit for auxiliary facility

☐ Modify licensed main facility   ☐ Modify licensed auxiliary facility

If purpose is to modify, indicate nature of change(s) by checking appropriate box(es), and specify the file number(s) of the authorization(s) affected:

☐ Antenna supporting-structure height   ☐ Effective radiated power

☐ Antenna height above average terrain   ☐ Frequency

☐ Antenna location   ☐ Antenna system

☐ Main Studio location   ☐ Other *(Summarize briefly)*

File Number(s) _____

1. Allocation:

Channel No.	Offset *(check one)*	Principal community to be served:			Zone *(check one)*
	☐ Plus	City	County	State	☐ I
	☐ Minus				☐ II
_____	☐ Zero				☐ III

2. Exact location of antenna:

(a) Specify address, town or city, county and state. If no address, specify distance and bearing to the nearest landmark.

(b) Geographical coordinates (to nearest second). If mounted on element of an AM array, specify coordinates of center of array. Otherwise, specify tower location. Specify South Latitude or East Longitude where applicable; otherwise, North Latitude and West Longitude will be presumed.

Latitude	°	'	"	Longitude	°	'	"

3. Is the supporting structure the same as that of another station(s) or proposed in another pending application(s)?   ☐ Yes   ☐ No

If Yes, give call letter(s) or file number(s) or both.   _____

If proposal involves a change in height of an existing structure, specify existing height above ground level, including antenna, all other appurtenances, and lighting, if any.   _____

▶ *Figure 10   (continued)*

4. Does the application propose to correct previous site coordinates?   ☐ Yes ☐ No
   If Yes, list old coordinates.

Latitude ° ′ ″	Longitude ° ′ ″

5. Has the FAA been notified of the proposed construction?   ☐ Yes ☐ No
   If Yes, give date and office where notice was filed and attach as an Exhibit a copy of FAA
   determination, if available.

   ☐ Exhibit No.

   Date _____ Office where filed_____

6. List all landing areas within 8 km of antenna site. Specify distance and bearing from structure to nearest point of
   the nearest runway.

Landing Area	Distance (km)	Bearing (degrees True)
(a) _____	_____	_____
(b) _____	_____	_____

7. (a) Elevation: *(to the nearest meter)*

   (1) of site above mean sea level;   _____ meters

   (2) of the top of supporting structure above ground (including antenna, all other   _____ meters
       appurtenances, and lighting, if any); and

   (3) of the top of supporting structure above mean sea level $[(aX1) + (aX2)]$.   _____ meters

   (b) Height of antenna radiation center: *(to the nearest meter)*

   (1) above ground;   _____ meters

   (2) above mean sea level $[(aX1) + (bX1)]$; and   _____ meters

   (3) above average terrain.   _____ meters

8. Attach as an Exhibit sketch(es) of the supporting structure, labelling all elevations required   ☐ Exhibit No.
   in Question 7 above, except item 7(b)(3). If mounted on an AM directional-array element,
   specify heights and orientations of all array towers, as well as location of TV radiator.

9. Maximum visual effective radiated power   _____ kW

▶ *Figure 10   (continued)*

FCC 574
August 1989

See reverse side for information regarding public burden estimate.

Federal Communications Commission

Approved by OMB
3060-0128
Expires 5/31/92

APPLICATION FOR PRIVATE LAND MOBILE AND GENERAL MOBILE RADIO SERVICES

IMPORTANT NOTICE: ALL applicants MUST include the PROPER FEE with their application and MUST FILE those applications at the PROPER LOCATION. Refer to the current fee publication or contact the Consumer Assistance Branch, Federal Communications Commission, Gettysburg, PA 17326 (717) 337-1212.

COMMISSION USE ONLY:

Page No. of

	1. Frequencies (MHz)	2. Station Class	3. No. of Units	4. Emission Designator	5. Output Power	6. E.R.P.	7. A.A.T.	8. Ground Elevation	9. Ant. Hgt. To Tip	10. Antenna Latitude	11. Antenna Longitude
A											

12. Number of Mobiles By Category:
Vehicular
Portable
Aircraft
Marine
Pagers

13. Area of Operation for Mobiles, Temporary, or Itinerant Stations
is ___ miles radius of station A,
or is ___ miles radius of coordinates:
Lat
Long
County
State
If not, please check ONE:
☐ Countywide
☐ Statewide
☐ Nationwide
☐ Other

	14. Station Address or Geographic Location	15. City	16. County	17. St.
A				
B				
C				
D				
E				
F				

18. Location of Primary Control Point (Include telephone number) and location of all Radio Control Stations with antenna under 20 ft. GMRS ONLY: List small control station locations and "FBA" followed by small base locations with antennas under 20 ft. (See instructions.)

19. Freq. Advisory Comm. No:

20. Radio Service:

21. Applicant/Licensee Name *(See Instructions)*:

22. Mailing Address *(Number & Street, P.O. Box or Rt. No)*:
ATTN:
ADDRESS:

23. City 24. State 25. ZIP Code

26. Will antenna be mounted on a structure with an existing antenna? If yes, give call sign and radio and radio service of existing licensee.

27. Provide description of the structure on which your antenna is mounted and the height above ground to the top of the structure. (See antenna figures 1-3 on reverse for samples.)

28. Give the name of the nearest aircraft landing area, and the distance and direction to the nearest runway.

	No	Yes	Call Sign	Radio Service		Structure Type	Structure Height Above Ground	Aircraft Landing Area Name	Distance (Miles)	Direction
A					A					
B					B					
C					C					
D					D					
E					E					
F					F					

29. Has notice of construction or alteration been filed with the FAA? If yes, give the date filed, the name under which filed, and the FAA office where filed.

	No	Yes	Date Filed	Name Under Which Filed	FAA Office Where Filed
A					
B					
C					
D					
E					
F					

30. Applicant Classification ☐ Individual ☐ Partnership ☐ Association ☐ Corporation ☐ Govt. Entity

31. Eligibility *(Describe Activity)*:

Give Rule Section

32. Application is for (check one): ☐ New Station ☐ Modification ☐ Assignment ☐ Reinstatement ☐ Renewal

33. Does application include the complete system? ☐ Yes ☐ No

34. Would a commission grant of your application be an action which may have a significant environmental effect as defined by Section 1.1307 of the Commission's Rules? If you answer yes, submit the statement as required by Sections 1.1308 and 1.1311. ☐ Yes ☐ No

35. Specification of item(s) modified, if applicable:

36. If other than new station, list call sign(s) of existing station(s) to be modified, combined or reinstated.

37. Individual completing this application form(s):

Telephone No. (   )

38. Supplemental Information for Trunked and Conventional Systems 806-824/851-869 MHz and 896-901/935-940 MHz frequency bands

Indicate Type of Applicant:
☐ a) Independent ☐ b) Commercial (SMRS entrepreneur)
☐ c) Community Repeater (Owner)
☐ d) SMRS user (Show SMRS licensee name and call sign and allocate your mobile loading)

Type of system: (Check One)
☐ a) Conventional. Specify the number of mobile units to be placed in operation at the time of grant:
☐ b) Trunked. Specify the number of trunked channels requested:

Frequency Band Requested: (Check One) ☐ a) 851-869 MHz ☐ b) 935-940 MHz

FOR COORDINATOR USE ONLY:

CERTIFICATION. READ CAREFULLY BEFORE SIGNING

1) If application is for a Land Mobile Service license, applicant certifies that a current copy of the requested radio service's rules will be obtained. Contact the United States Government Printing Office, Washington, DC 20402 (202) 783-3238.
2) Applicant waives any claim to the use of any particular frequency regardless of prior use by license or otherwise.
3) Applicant will have unlimited access to the radio equipment and will control access to exclude unauthorized persons.
4) Neither applicant nor any member thereof is a foreign government or representative thereof.
5) Applicant certifies that all statements made in this application and attachments are true, complete, correct and made in good faith.
6) Applicant certifies that the signature is that of the individual, or partner, or officer or duly authorized employee of a corporation, or officer who is a member of an unincorporated association, or appropriate elected or appointed official on behalf of a governmental entity.

WILLFUL FALSE STATEMENTS MADE ON THIS FORM ARE PUNISHABLE BY FINE AND/OR IMPRISONMENT. U.S. CODE TITLE 18, SECTION 1001

Typed/Printed Name: Telephone No: (   )

Signature: 'MUST BE ORIGINAL' Date:

▶ ***Figure 11*** *Form 574. Applications for private land mobile and general mobile radio services.*

NOTICE TO INDIVIDUALS REQUIRED BY PRIVACY ACT OF 1974
AND THE PAPERWORK REDUCTION ACT OF 1980

Sections 301, 303, and 308 of the Communications Act of 1934, as amended (licensing powers) authorized the FCC to request the information on this application. The purpose of the information is to determine your eligibility for a license. The information will be used by FCC staff to evaluate the application, to determine station location, to provide information for enforcement and rulemaking proceedings and to maintain a current inventory of licensees. No license can be granted unless all information requested is provided. Your reponse is required to obtain this authorization.

Public reporting burden for this collection of information is estimated to range from fifteen minutes to six hours per response, including time for reviewing instructions, searching existing data sources, gathering and maintaining the data needed, and completing and reviewing the collection of information. Send comments regarding the burden estimate or any other aspect of this collection of information, including suggestions for reducing this burden, to Federal Communications Commission, Office of Managing Director, Washington, DC 20554, and to the Office of Information and Regulatory Affairs, Office of Management and Budget, Paperwork Reduction Project (3060-0128), Washington, DC 20503.

USE THE SAMPLE INFORMATION SHOWN BELOW TO ASSIST YOU WITH ANSWERING ITEMS 8, 9, AND 27 ON REVERSE SIDE.

ANTENNA ARRANGEMENTS (NOT TO SCALE)

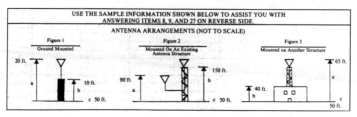

▽ = antenna
a = height above ground to tip of proposed antenna (item 9)
b = height above ground to top of supporting structure (item 27)
c = ground elevation above mean sea level (item 8)

For these figures, items 8, 9 and 27 would be completed as follows (See samples below):

	Item 8	Item 9	Item 27	
	Ground Elevation	Antenna Height To Tip	Structure Type	Structure Height Above Ground
Figure #1	50	20	Pole	10
Figure #2	50	80	Tower	150
Figure #3	50	65	Building	40

▶ *Figure 11* *(continued)*

Codes and Names of the Radio Services

Part I All frequencies except 800 and 900 MHz Bands

	Industrial			Motor Carrier
(IB)	Business		(LI)	Interurban Passenger
(IF)	Forest Products		(LJ)	Interurban Property
(IM)	Motion Picture		(LU)	Urban Passenger
(IP)	Petroleum		(LV)	Urban Property
(IS)	Special Industrial			
(IT)	Telephone Maintenance			**Public Safety**
(IW)	Power		(PF)	Fire
(IX)	Manufacturers		(PH)	Highway Maintenance
(IY)	Relay Press		(PL)	Local Government
			(PP)	Police
	Land Transportation		(PO)	Forestry Conservation
(LA)	Automobile Emergency			
(LR)	Railroad		(PS)	**Special Emergency**
(LX)	Taxicab			
			(RS)	**Radiolocation**
			(ZA)	**General Mobile**

Part II 806–821/851–866 MHz Bands

	Conventional	Trunked
Business	(GB)	(YB)
Industrial/Land Transportation	(GO)	(YO)
Public Safety/Special Emergency	(GP)	(YP)
Commercial (SMRS)	(GX)	(YX)

Part III 821–824/866–869 MHz Bands

	Conventional	Trunked
Public Safety/Special Emergency National Plan	(GF)	(YF)

Part IV 929–930 MHz Band

	Conventional
Private carrier paging systems (PCPS) (All other applicants use the code for the radio service in which eligibility is claimed. See Part I of this table)	(GS)

Part V 896–901/935–940 MHz Bands

	Conventional	Trunked
Business	(GU)	(YU)
Industrial/Land Transportation	(GI)	(YI)
Public Safety/Special Emergency	(GA)	(YA)
Commercial (SMRS)	(GR)	(YS)

▶ *Figure 11   (continued)*

Federal Communications Commission
Washington, D.C. 20554

Approved by OMB
3060-0113
Expires 9/30/90

BROADCAST EQUAL EMPLOYMENT OPPORTUNITY PROGRAM REPORT

(To be filed with broadcast license renewal application)

(For FCC Use Only)

Call Letters .... _____

Code No.

Name of Licensee .... _____

City and State which station
    is licensed to serve .... _____

TYPE OF BROADCAST STATION (Check one)

Commercial Broadcast Station

Noncommercial Broadcast Station

☐ AM

☐ TV

☐ Educational Radio

☐ FM

☐ Low Power TV

☐ Educational TV

☐ Combined AM & FM
   in same area

☐ International

SEND NOTICES AND COMMUNICATIONS TO THE FOLLOWING NAMED PERSON AT THE ADDRESS INDICATED BELOW:

Name	Street Address		
City	State	ZIP Code	Telephone No. ( )

FILING INSTRUCTIONS

Broadcast station licensees are required to afford equal opportunity to all qualified persons and to refrain from discriminating in employment and related benefits on the basis of race, color, national origin, and sex. See Section 73.2080 of the Commission's Rules. Pursuant to these requirements, a license renewal applicant who employs five or more full-time station employees must file a report of its activities to ensure equal employment opportunity for women and minority groups (that is, Blacks not of Hispanic origin, Asians or Pacific Islanders, American Indians or Alaskan Natives, and Hispanics). If minority group representation in the available labor force is less than five percent (in the aggregate), equal employment opportunity (EEO) program information for minority group members need not be filed. However, EEO program information must be filed for women since they comprise a significant percentage of virtually all area labor forces. If an applicant employs fewer than five full-time employees, no equal employment opportunity activity information need be filed.

A copy of this report must be kept in the station's public file. These actions are required to obtain license renewal. Failure to meet these requirements may result in license renewal being delayed or denied. These requirements are contained in Section 73.2080 of the FCC Rules (47 CFR 73.2080), and are authorized by the Communications Act of 1934, as amended.

☐ If your station employs fewer than five full-time employees, check the box at left, complete the certification below, return the form to the FCC, and place a copy in your station's public file. You do not have to complete the rest of the form.

If your station employs five or more full-time employees, you must complete all of this form and follow all instructions.

☐ If minority group representation in the available labor force is less than 5 percent (in the aggregate) and you choose not to file EEO program information for minority groups, check the box at left and complete the rest of this form with only the information for your program directed towards women.

▶ *Figure 12   Broadcast Equal Employment Opportunity program report.*

☐ Recruit prospective employees from educational institutions, including area schools and colleges with minority and women enrollments. Educational institutions contacted for recruitment purposes during the past 12 months and the number of minority and/or women referrals are:

Educational Institution	Number of Referrals	
	Minority	Women
_____	_____	_____
_____	_____	_____

☐ Contact a variety of minority and women's organizations to encourage the referral of qualified minority and women applicants whenever job vacancies occur. Examples of such organizations contacted during the past 12 months are:

Organization	Number of Referrals	
	Minority	Women
_____	_____	_____
_____	_____	_____
_____	_____	_____
_____	_____	_____

☐ We encourage present employees to refer qualified minority and women candidates for job openings. The number of minority and/or women referrals are:

Minority _____   Women _____

☐ Other (specify) and the number of minority and/or women referrals are:

Minority _____   Women _____

IV. JOB HIRES

A broadcast station must consider applicants for job openings on a nondiscriminatory basis. Further, to assure that qualified minorities and women are given due consideration for available positions, it must make efforts to encourage them to apply for job openings.

During the twelve-month period prior to filing this application beginning (Month-Day-Year) _____ and ending (Month-Day-Year), _____ we hired:

Total hires _____   Minorities _____   Women _____

During this period, for positions in the upper four job categories, we hired:

Total hires, upper _____   Minorities _____   Women _____
four categories

V. PROMOTIONS

A broadcast station must promote individuals on a nondiscriminatory basis. Further, to assure that qualified minorities and women are given due consideration for promotional opportunities, it must make efforts to encourage them to qualify and apply for advancement.

During the twelve-month period prior to filing this application beginning (Month-Day-Year) _____ and ending (Month-Day-Year) _____ we promoted:

Total promotions _____   Minorities _____   Women _____

During this period, in the upper four job categories, we promoted:

Total promotions, upper _____   Minorities _____   Women _____
four categories

VI. AVAILABLE LABOR FORCE

A broadcast station must evaluate its employment profile and job turnover against the availability of minorities and women in the relevant labor market. The FCC will use labor force data for the MSA in which your station is located, or county data if the station is not located in an MSA, to evaluate your station's equal employment efforts. If you use these data in your evaluation, you need not submit them to the FCC.

▶ *Figure 12* *(continued)*

Approved by OMB
3060-0010
Expires 04/30/89

CERTIFICATION

United States of America
Federal Communications Commission
Washington, D. C. 20554

I certify that I am _____

(*Official title, see Instruction 1*)

of _____

(*Exact legal title or name of respondent*)

Ownership Report

NOTE: Before filling out this form, read attached instructions

that I have examined this Report, that to the best of my knowledge and belief, all statements in the Report are true, correct and complete.

Section 310(d) of the Communications Act of 1934 requires that consent of the Commission must be obtained prior to the assignment or transfer of control of a station license or construction permit. This form may not be used to report or request an assignment of license/permit or transfer of control (except to report an assignment of license/permit or transfer of control made pursuant to prior Commission consent).

(*Date of certification must be within 60 days of the date shown in Item 1 and in no event prior to Item 1 date*):

_____ _____, 19 _____

(*Signature*)   (*Date*)

---

1. All of the information furnished in this Report is accurate as of

_____ , 19 _____ .

(*Date must comply with Section 73.3615(a), i.e., information must be current within 60 days of the filing of this report, when 1(a) below is checked.*)

Telephone No. of respondent (*include area code*):

Any person who willfully makes false statements on this report can be punished by fine or imprisonment. U.S. Code, Title 18, Section 1001.

This report is filed pursuant to Instruction (*check one*)

Name and Post Office Address of respondent:

1(a) ☐ Annual   1(b) ☐ Transfer of Control or Assignment of License   1(c) ☐ Other

for the following stations:

Call Letters	Location	Class of service

4. Name of entity, if other than licensee or permittee, for which report is filed (*see Instruction 3*):

2. Give the name of any corporation or other entity for whom a separate Report is filed due to its interest in the subject licensee (*See Instruction 3*):

5. Respondent is:

☐ Sole Proprietorship

☐ For-profit corporation

☐ Not-for-profit corporation

3. Show the attributable interests in any other broadcast station of the respondent. Also, show any interest of the respondent, whether or not attributable, which is 5% or more of the ownership of any other broadcast station or any newspaper or CATV entity in the same market or with overlapping signals in the same broadcast service, as described in Sections 73.3555 and 76.501 of the Commission's Rules.

☐ General Partnership

☐ Limited Partnership

☐ Other: _____

If a limited partnership, is certification statement included as in Instruction 4?

☐ Yes   ☐ No

▶ *Figure 13 Ownership report.*

8.   List officers, directors, cognizable stockholders and partners. Use one column for each individual or entity. Attach additional pages, if necessary. See Instructions 4, 5, and 6.

Line   *(Read carefully - The numbered items below refer to line numbers in the following table.)*

1. Name and residence of officer, director, cognizable stockholder or partner (if other than individual also show name, address and citizenship of natural person authorized to vote the stock). List officers first, then directors and, thereafter, remaining stockholders and partners.

2. Citizenship.

3. Office or directorship held.

4. Number of shares or nature of partnership interest.

5. Number of votes.

6. Percentage of votes.

7. Other existing attributable interests in any other broadcast station, including nature and size of such interest.

8. All other ownership interests of 5% or more (whether or not attributable), as well as any corporate officership or directorship in broadcast, cable, or newspaper entities in the same market or with overlapping signals in the same broadcast service, as described in Sections 73.3555 and 76.501 of the Commission's Rules, including the nature and size of such interests and the position held.

1	(a)	(b)	(c)
2			
3			
4			
5			
6			
7			
8			

**FCC NOTICE TO INDIVIDUALS REQUIRED BY THE PRIVACY ACT
AND THE PAPERWORK REDUCTION ACT**

The solicitation of personal information requested in this Report is authorized by the Communications Act of 1934, as amended. The principal purpose for which the information will be used is to assess compliance with the Commission's multiple ownership restrictions. The staff, consisting variously of attorneys and examiners, will use the information to determine such compliance. If all the information requested is not provided, processing may be delayed while a request is made to provide the missing information. Accordingly, every effort should be made to provide all necessary information. Your response is required to retain your authorization.

**THE FOREGOING NOTICE IS REQUIRED BY THE PRIVACY ACT OF 1974, P.L. 95-579, DECEMBER 31, 1974,
5 U.S.C. 552(d)(3) AND THE PAPERWORK REDUCTION ACT P.L. 96-511, DECEMBER 11, 1980, 44 U.S.C. 3507.**

▶   *Figure 13   (continued)*

Call Signs

The need for station identification is a worldwide problem. There are literally millions of broadcast and non-broadcast stations throughout the world. International agreements, beginning in 1927, have divided the alphabet among different countries so that the first call sign letter identifies the nationality of the station. For example, the United States—which has 85 different kinds of radio services for land, sea, air, and space communications—is assigned three letters exclusively, *K*, *W*, and *N*, and shares the letter *A* with some other countries. The Communications Act gives the FCC authority to assign call letters to all stations under its jurisdiction. A and N are designated for government use: A for the Army and Air Force stations, and N to the Navy and Coast Guard. The letters K and W are assigned to all domestic stations, both government and non-government.

Broadcast stations west of the Mississippi River and in U.S. territories and possessions are assigned the letter K, with the letter W going to stations east of the Mississippi River. In the early days of radio, before the call-letter allocation plan went into effect, most stations were in the east, and many of them retained their original K call sign. At first only three letters were assigned; as the number of stations grew, call signs grew to four letters.

Stations may request their own call letters, provided that call sign is not already taken. In some instances where the requested call sign is very similar to that of an already existing station in the same area—for example, WKCB and WKCP—it may be denied. Call letters may reflect slogans or other special identifications. For example, WNYC, New York, is the New York City municipal station; WIOD, Miami, stands for "Wonderful Isle Of Dreams"; WTOP, Washington, D.C., means "Top of the Dial"; and WGCD, Chester, South Carolina, is "Wonderful Guernsey Center of Dixie." Many call letters reflect the owner's name or initials. Where the same licensee owns an AM and another station in the same area, different signs are not necessary, but the class of service is added to the identification, such as WKXN and WKXN-FM, or WKXN-TV. Some foreign stations do not use call signs, but identify the station by the city of origin. Amateur or ham radio station call signs are assigned by an automatic system depending on the operator's license class and mailing address.

Renewals

Part of the licensing process is the renewal stage. As noted earlier, until deregulation of the renewal procedure in 1983 stations had to file extensive and detailed application forms justifying how they had served the public interest of their communities during the preceding 3 years of the license period. Deregulation extended license terms to 5 years for television and 7 years for radio. As of 1991, renewal applications continue to be filed on sheets the size of large postcards, with a minimum of information required (see Figure 14).

A renewal applicant must broadcast announcements to the public of its filing of the renewal application. The FCC specifies the frequency and timing of such announcements. The FCC requires a *pre-filing announcement*, with a sample text advising the public that the station's license will expire, that the application for renewal may be examined at the station, and inviting the public to advise the FCC whether it believes the station has or has not operated in the public interest. It also

Federal Communications Commission
Washington, D.C. 20554

APPLICATION FOR RENEWAL OF LICENSE FOR
COMMERCIAL AND NONCOMMERCIAL AM, FM OR TV BROADCAST STATION

Approved by OMB
3060-0110
Expires 5/31/91

For Commission Fee Use Only		For Applicant Fee Use Only

FEE NO:

FEE TYPE:

FEE AMT:

ID SEQ:

For Commission Use Only: File No.

Is a fee submitted with this application?   ☐ Yes ☐ No

If No, indicate reason therefor (check one box):
☐ Nonfeeable application

Fee Exempt (See 47 C.F.R. Section 1.1112)

☐ Noncommercial educational licensee

☐ Governmental entity

1. Name of Applicant

Mailing Address

City	State	ZIP Code

2. This application is for: ☐ AM ☐ FM ☐ TV

(a) Call Letters: (b) Principal Community:
City   State

3. Attach as Exhibit No. _____ an identification of any FM booster or TV booster station for which renewal of license is also requested.

4. Have the following reports been filed with the Commission:

(a) The Broadcast Station Annual Employment Reports (FCC Form 395-B) as required by 47 C.F.R. Section 73.3612?   ☐ Yes ☐ No

If No, attach as Exhibit No. _____ an explanation.

(b) The applicant's Ownership Report (FCC Form 323 or 323-E) as required by 47 C.F.R. Section 73.3615?   ☐ Yes ☐ No

If No, give the following information:
Date last ownership report was filed   _____
Call letters of station for which it was filed _____

FCC 303-S
May 1988

---

5. Is the applicant in compliance with the provisions of Section 310 of the Communications Act of 1934, as amended, relating to interests of aliens and foreign governments?   ☐ Yes ☐ No

If No, attach as Exhibit No. _____ an explanation.

6. Since the filing of the applicant's last renewal application for this station or other major application, has an adverse finding been made or final action been taken by any court or administrative body with respect to the applicant or parties to the application in a civil or criminal proceeding, brought under the provisions of any law relating to the following: any felony; broadcast related antitrust or unfair competition; criminal fraud or fraud before another governmental unit; or discrimination?   ☐ Yes ☐ No

If Yes, attach as Exhibit No. _____ a full description of the persons and matters involved, including an identification of the court or administrative body and the proceeding (by dates and file numbers) and the disposition of the litigation.

7. Would a Commission grant of this application come within 47 C.F.R. Section 1.1307, such that it may have a significant environmental impact?   ☐ Yes ☐ No

If Yes, attach as Exhibit No. _____ an Environmental Assessment required by 47 C.F.R. Section 1.1311.

If No, explain briefly why not.

8. Has the applicant placed in its station's public inspection file at the appropriate times the documentation required by 47 C.F.R. Sections 73.3526 or 73.3527?   ☐ Yes ☐ No

If No, attach as Exhibit No. _____ a complete statement of explanation.

The APPLICANT hereby waives any claim to the use of any particular frequency or of the electromagnetic spectrum as against the regulatory power of the United States because of the previous use of the same, whether by license or otherwise, and requests an authorization in accordance with this application. (See Section 304 of the Communications Act of 1934, as amended.)

The APPLICANT acknowledges that all the statements made in this application and attached exhibits are considered material representations and that all the exhibits are a material part hereof and are incorporated herein as set out in full in the application.

CERTIFICATION: I certify that the statements in this application are true, complete, and correct to the best of my knowledge and belief, and are made in good faith.

Name	Signature
Title	Date

WILLFUL FALSE STATEMENTS MADE ON THIS FORM ARE PUNISHABLE BY FINE AND IMPRISONMENT. U.S. CODE, TITLE 18, SECTION 1001.

▶ *Figure 14   Application for Renewal of License for Commercial and Noncommercial AM, FM, or TV Broadcast Station. Initial information and items 1–4 are on one side of the renewal form, items 5–8 and the certification on the other.*

requires a *post-filing announcement* after the renewal form has been submitted, again inviting the public to send comments to the FCC. Here is the sample text for the pre-filing announcement; the text for the post-filing announcement is virtually identical.

> On (date of last renewal grant), (station's call letters) was granted a license by the Federal Communications Commission to serve the public interest as public trustee until (expiration date).

> Our license will expire on (date). We must file an application for license renewal with the FCC on (date four calendar months prior to expiration date). When filed, a copy of this application will be available for public inspection during our regular business hours. It contains information concerning this station's performance during the last license period (period of time covered by the application).

> Individuals who wish to advise the FCC of facts relating to our renewal application and to whether this station has operated in the public interest should file comments and petitions with the Commission by (date first day of calendar month prior to month of expiration).

> Further information concerning the Commission's broadcast license renewal process is available at (address of location of the station's public inspection file) or may be obtained from the FCC, Washington, D.C. 20554.

An examination of the license renewal form shows that the key concerns of the FCC are equal employment opportunity, foreign ownership, misconduct such as criminal activity or violations of certain U.S. statutes, environmental impact, and ownership. The station itself represents whether it has operated in the public interest; the FCC does not examine its programming or other services to the community.

PUBLIC PARTICIPATION

Although the FCC's deregulatory policies in the late 1970s and throughout the 1980s sharply curtailed the public's participation in FCC operations and actions, the Commission in the 1990s continued to state its interest in having consumer input into FCC policies and procedures. For example, it endorsed National Consumer's Week in April, 1990, citing "the vital role that consumers play in the marketplace," and offered several events at FCC headquarters and at field offices on consumer issues. One such issue dealt with the individual's rights of privacy in an age when communication technologies make it possible to obtain and use personal information. Another issue was how a member of the public could file a telephone complaint with the FCC regarding any of the services—telephone, broadcasting, private radio, and others—under FCC jurisdiction, and what happens to complaints after they are received. The FCC has issued bulletins on how consumers may make their concerns known on specific communications services. One such bulletin deals with an area with a high incidence of complaints, cable television. The bulletin states the areas not regulated by the FCC, thus making complaints to the FCC about them useless, and notes to what specific authority complaints may

PUBLIC NOTICE

Federal Communications Commission
1919 M Street, N.W.
Washington, D.C. 20554 2603
2

News media information 202/632-5050. Recorded listing of releases and texts 202/632-0002.

REPORT NO. 32

OPEN PROCEEDINGS

April 9, 1990

The following is a listing of current FCC Notices of Proposed Rulemaking (NPRM) and Notices of Inquiry (NOI) open for public comment. This listing covers all outstanding NPRMs and NOIs, except routine petitions to amend the Table of Allotments, as of April 5, 1990. For additional information contact Martha Contee, Consumer Assistance and Small Business Division (202) 632-7000.

DOCKET #	NOI/NPRM: DATE RELEASED: FCC/MIMEO #	SUBJECT/ISSUE	DEADLINE TO FILE COMMENTS/REPLIES	STAFF CONTACT PHONE (202)
GEN 88-441	FNOI: 12/11/89 FCC 89-326	RE: Advanced Technologies for the Public Safety Radio Service	May 15, 1990* Jul. 16, 1990	Marty Liebman 632-6497
	ORDER: 2/6/90 DA 90-116	**SUMMARY:** This Further Notice addresses the concerns that the Commission has about public safety interoperability and future spectrum requirements for emerging digital radio technology.		
GEN 89-554	NOI: 12/13/89 FCC 89-329	RE: Preparation for the International Tele-communication Union World Administrative Radio Conference **SUMMARY:** In this Notice the Commission is inviting comments concerning its preparation for the 1992 International Telecommunication World Administrative Radio Conference (WARC-92), on frequency allocations in certain parts of the spectrum. WARC-92 is expected to address a variety of allocation issues resulting from three prior WARCs on High Frequency broadcasting, mobile services and space radio communication services.	Feb. 16, 1990* Mar. 16, 1990	William Torak 632-7025

Figure 15 *Public Notices.*

CC 90-75	NPRM: 3/15/90 FCC 90-77	RE: Amendment of Part 22 of the Commission's Rules **SUMMARY:** The Commission is proposing to amend Part 22 of its rules to require all potential applicants for Public Land Mobile Service facilities to coordinate their engineering proposals with existing licensees and applicants whose own systems or proposed systems might experience interference from the new proposal.	May 07, 1990 May 22, 1990	Gerald M. Zuckerman 632-6450
MM 89-494	NOI: 11/20/89 FCC 89-307 ORDER: 1/12/90 DA 90-24 ORDER: 3/19/90 DA 90-24	RE: Enforcement of Prohibitions Against Broadcast Indecency in 18 U.S.C. S.1464 **SUMMARY:** In this Notice of Inquiry the Commission is soliciting comments to compile a record related to a judicial determination of the validity of a 24-hour ban on indecent programming. The ban was ordered by the Commission in compliance with a congressional directive, and stayed by the U.S. Court of Appeals D.C. Circuit.	Feb. 20, 1990 Apr. 19, 1990 **	Marilyn Mohrman-Gillis 632-7792
MM 89-600	NOI: 12/29/89 FCC 89-345	RE: Cable Competition and Rates **SUMMARY:** The Commission is soliciting comments to compile a comprehensive study of the cable industry's operations under the 1984 Cable Act. Upon completion of the study the Commission will submit to Congress a report analyzing the effect on the video services marketplace of substituting market forces for cable rate regulations.	Mar. 1, 1990* Apr. 2, 1990	Scott Roberts 632-6302 OR David Horowitz 632-7792

▶ *Figure 15*   *(continued)*

		RE / SUMMARY	Dates	Contact
PR 90-26	NPRM: 2/5/90 FCC 90-33 PN: 3/30/90 DA 90-528	RE: Amendment of the Maritime Service Rules SUMMARY: The Commission is proposing to amend Part 80 of its Rules to require that VHF ship station transmitters automatically cease operation after a predetermined period of operation. Under the proposed rules, VHF ship station transmitters would be equipped with: (1) an automatically timing device that deactivates the transmission after an uninterrupted transmission period of more than three minutes; and (2) a device that provides an in-dication that the timer has deactivated the transmitter.	May 10, 1990 May 25, 1990	George R. Dillon 632-7175
PR 90-27	NPRM: 2/1/90 FCC 90-32	RE: Amendment of the Maritime Service Rules SUMMARY: The Commission is proposing to allow both commercial and noncommercial communications on VHF Channels 67 and 72 in the Puget Sound Area.	Mar. 23, 1990 * Apr. 09, 1990	Robert DeYoung 632-7175
PR 90-34	NPRM: 2/28/90 FCC 90-38	RE: Amendment of Part 90 of the Commission's Rules SUMMARY: The Commission is proposing to eliminate the need for waivers for short spacing of Specialized Mobile Radio systems where (1) all potentially affected co-channel licensees agree to the proposed location and (2) all concurring licensees' co-channel systems are constructed and fully operational.	Apr. 23, 1990 May 08, 1990	Irene Bleiweiss 634-2443
PR 90-55	NPRM: 2/16/90 FCC 90-62	RE: Amendment of Part 97 of the Commission's Rules SUMMARY: The Commission is proposing to establish a new class of amateur operator license that would not require the applicant to prove that he or she can send and receive manual Morse code telegraphy messages.	Aug. 06, 1990 Sept. 07, 1990	Maurice J. DePont 632-4964

▶ *Figure 15*   (continued)

MM 90-4	NPRM: 1/22/90 FCC 90-12	RE: Cable Effective Competition Reexamination **SUMMARY:** The Commission is proposing to reexamine the rules regarding the regulation of basic cable service rates. Under the 1984 Cable Policy Act, the Commission adopted a "three signal standard" for determining when cable communities were not subject to "effective competition," so that local franchising authorities could regulate basic service rates. Now the Commission believes that changed circumstances in the video marketplace warrant review of the three signal standard, and is soliciting comments on what, if any, reforms are needed.	Arp. 6, 1990 May 7, 1990	Marcia Glauberman 632-6302
MM 90-162	NPRM: 3/14/90 FCC 90-100	RE: Evaluation of the Syndication and Financial Interest Rules **SUMMARY:** This Notice begins a new rulemaking proceeding to evaluate the financial interest and syndication rules. The Commission believes the analysis of potential options would be sharpened by a record focusing on existing and future marketplace realities rather than one including voluminous outdated materials.	June 14, 1990 Aug. 01, 1990	David Solomon 632-6990 and William Johnson 632-6460
MM 88-140	NPRM: 3/28/90 FCC 90-93	RE: Amendment of Part 74 of the Commission's Rules **SUMMARY:** The Commission is proposing to revise and clarify FM Translator rules, including new rules for ownership and financial support of translators; methods for selecting among translator applications; use of commercial, noncommercial and technical requirements for translators.	June 15, 1990 July 16, 1990	Tatsu Kondo 632-6302

▶ *Figure 15* *(continued)*

be made concerning rates, programming, access, and technical standards—the principal consumer concerns.

In addition, the FCC issues regular Public Notices of matters before the Commission on which public comments may be submitted. While few individual members of the public see or are interested in these Public Notices, they are seen by the various national, state, and local organizations representing interested consumers, and by the small groups of consumers who may be affected by a specific local matter before the FCC. Many of the larger national groups, such as Action for Children's Television, National Black Media Coalition, and National Organization for Women Media Task Force file comments and petitions on a regular basis.

As an example of the kinds of matters the Commission deals with and which are open for public comment, Figure 15 shows part of a public notice on "open proceedings."

Filing a Complaint

The principal complaints relate, as one might expect, to the two areas of service used most by individual members of the public: telephone and broadcasting. An analysis of the procedure for filing a complaint against a broadcaster will serve as a generic description—and a guideline—of the basic process for filing a complaint on any service.

The FCC has a limited enforcement staff. It is virtually impossible for the FCC to seek out violations of the Rules and Regulations. Some years ago Commissioner Nicholas Johnson pointed out that the Commission had three field investigators and FCC regulations required field investigators to go around in pairs. Complaints from the public (and from other competitor stations in the community) usually are the first notices the FCC has of any violations, including those related to technical operations.

The FCC can act only on complaints relating to violations of provisions of the Communications Act or of the FCC's Rules and Regulations. On the one hand, it cannot act on complaints that programs have too much violence, because there is no provision in the Act or in the Rules regarding violence. On the other hand, both the Act and the Rules contain provisions regarding indecency, and with proof of violations of such provisions, the FCC can take action. Some years ago the FCC was picketed for several days by a group objecting to the inclusion of the issue of abortion in an episode of the CBS sitcom, "Maude." In fact, such content was outside the purview of the FCC; the picketing group would have been more on the mark had they directed their concern to the Washington offices of CBS, just a couple of blocks away.

Process   The first step is to complain to the station itself. Sometimes the station is able to explain the situation to the complainant's satisfaction; sometimes it recognizes the merits of the complaint and takes corrective action. If the complainant (whether an individual or a group) is not satisfied with the station's response, the complaint then may be taken directly to the FCC. It is important to act on concerns as soon as possible, so that the FCC receives the complaint as quickly as feasible following the offending programming or other station action. It is important that the FCC receive as many facts as possible, including full information on the complainant (name and address) and the station (location and call letters), and the specific nature of the station's action that prompts the complaint, including the name of the program and the day and time it was broadcast (if the complaint relates to program content). Names of station personnel contacted and copies of any correspondence or notes on phone conversations should be included. Ordinarily, the FCC prefers to have complaints in writing. Sometimes, the critical nature of the alleged violation—for example, insufficient lights on an antenna tower, creating an immediate air safety hazard, or spurious emission from a station's transmitter, causing interference with other stations or communication services—requires a phone call, with follow-up in writing, if necessary. As noted in Chapter 3, complaints and investigations offices exist for all the FCC-regulated services.

If the FCC determines that the facts submitted indicate a violation of the Rules, it may investigate by first contacting the station (or other violator in other services) and then, if warranted, order a field investigation. In technical matters the Field Operations Bureau usually asks the appropriate field office to participate. If the investigation concludes that the Communications Act or the FCC Rules and Regulations have been violated, the FCC may impose sanctions that range from a letter of admonition, to a fine as high as $250,000, to a hearing to determine if the station's license should be revoked. If the investigation shows that, in the opinion of the FCC, there has been no legal violation, the complainant is sent a letter explaining the FCC's determination.

In some cases, when an individual or group's complaint has been denied by the FCC, the complainant waits until the station's license is up for renewal and then files a Petition to Deny that renewal. From time to time, particularly before deregulation limited the public's participation in the renewal process, licensees have been forced to make corrections in programming or other activities before a full renewal term is granted, and in some instances licenses have been revoked.

From a pragmatic standpoint, the most effective method the consumer has of affecting FCC policies and actions is, as noted earlier, through the political process. The most likely way of achieving results is by getting one or more members of the House or Senate committees dealing with communications and appropriations (see Chapter 1) to look into the given matter.

6

Public Information

Prior to deregulation in the 1980s the FCC public affairs office published and disseminated a large number of bulletins to the public on the operations of the FCC, methods of public participation in that process, and overviews of various regulated services. Many of these bulletins, such as *Public Radio* and *Public Television*, have been discontinued, and the number of publications available free to the public has greatly diminished. Nevertheless, the FCC maintains a Consumer Assistance and Small Business Division that provides those materials that are available and assists the public—whether individual consumer, industry representative, or researcher— in locating information on the FCC's policies and programs. It publishes a booklet entitled *Information Seekers Guide: How to Find Information at the Federal Communications Commission*. Assistance in securing copies of or access to any of the FCC documents noted in this book may be obtained by writing to the Consumer Assistance and Small Business Division, Office of Public Affairs, Room 254, FCC, 1919 M Street, Washington, D.C. 20554, or by phoning 202-632-7000.

INFORMATION BULLETINS

In 1990 any member of the public could obtain directly from the FCC one copy of the following bulletins (duplicate copies require a reproduction fee):

How to Apply for a Broadcast Station
Mass Media Services
The FCC in Brief
Radio Station and Other Lists
Private Radio Services
Evolution of Wire and Radio Communications
How the Rules Are Made
Station Identification and Call Signs
Frequency Allocation
Memo to All Young People Interested in Radio
Cable Television

In addition, shorter bulletins or fact sheets may be obtained on *Low- Power Television (LPTV)*, *Multipoint-Distribution Service (MDS)*, *Instructional Television Fixed Service (ITFS)*, *Direct Broadcast Service (DBS)*, *Cellular Radio*, *Satellite Program Scrambling*, *Indecency/Obscenity*, *Ex parte*, *Tax Certificates*, *Dial-a-Porn*,

FCC Fee Information, Specialized Mobile Radio Service (SMRS), and *Syndex.*. From time to time the FCC publishes a more extensive public document on a key issue. An example is the 1984 88-page edition of *The Law of Political Broadcasting and Cablecasting: A Political Primer*. A comprehensive summary of recent FCC activities is the *Annual Report/Fiscal Year 19—*, which gives an overview of the issues dealt with that year by the Commission's various bureaus and offices.

PUBLIC NOTICES AND RELEASES

Press releases, public notices, texts of decisions, notices of proposed rule making, and similar materials are not available for general distribution. These may be read in the News Media Division of the Office of Public Affairs and copied on coin-operated duplicating machines. They may be purchased through one of the duplicating/distribution companies authorized by the FCC, and received either in traditional paper hard copy form or electronically to computers. The FCC publishes a *Daily Digest*, a compilation of all of the notices issued each day by the FCC, available in limited number on a first come–first served basis at the Public Affairs Office.

Video and audio tape recordings of the Commission's open meetings are available to the public through a private contractor for specified fees. Similarly, transcripts of FCC hearings may be obtained through a reporting service for a fee. The Consumer Assistance Division can provide information on these service companies.

FCC Record

This is published every two weeks with a comprehensive listing of all FCC actions during that period. It is available at an annual subscription rate from the Government Printing Office, Superintendent of Documents, Washington, D.C. 20402; 202-783-3238.

Federal Register

This daily publication of all federal actions includes summaries of FCC proposed rule makings and policy decisions, and full texts of final rules. The *Federal Register* is a good continuing source for researchers and may be found in larger university and community libraries, as well as at the FCC.

FCC RULES AND REGULATIONS

These are *not* obtainable from the FCC. They must be purchased from the U. S. Government Printing Office. The Rules and Regulations, which are in Title 47 of the Code of Federal Regulations, comes in five volumes, as described in Chapter I of this book. Volume I includes FCC organization, practice and procedure, commercial radio operation, experimental radio services other than broadcast, markings and lighting of antenna structures, radio frequency devices, frequency allocations and treaty matters, and industrial, scientific and medical equipment. Volumes II and III deal with Common Carrier services. Volume IV covers Mass Media services. Volume V includes Private Radio and Direct Broadcast Satellite. The cost for each volume is available from the Superintendent of Documents.

FCC LIBRARY

The FCC Library contains all of the FCC's past records, including annual reports, the Federal Register, proposed and enacted congressional legislation pertaining to communications, court decisions, and a comprehensive set of legal, technical, and other books and reference material, as well as trade journals and scholarly magazines pertaining to communications.

DOCKETS REFERENCE ROOM

Administered by the Office of the Secretary, this room contains materials of an active nature, such as petitions for rule making, dockets for adjudicatory and rule-making proceedings, and docket correspondence files. All dockets are maintained for a year following final disposition of a case, and after that are sent to the Federal Record Center in Suitland, Maryland. They may be obtained for viewing at the FCC with about two weeks notice. Materials dated prior to 1970 are filed with the National Archives and cannot be returned to the FCC, but can be seen only through arrangements with the Archives.

The Office of the Secretary also has an Agenda Branch that maintains files of all FCC actions, including minutes of the proceedings of Commission meetings and the final Commission vote. A Publication Branch has information concerning the history of the FCC's rules.

BUREAU FILES

Applications and materials relating to individual communication services are maintained by the Bureau responsible for the specific service.

Mass Media Bureau

The Mass Media public reference room contains station license files; new applications, assignments and transfers; engineering files; construction permits; it includes LPTV and ITFS. The Auxiliary Services Branch has documents on FM booster stations; international communication; intercity relays; remote pickups; translator relays; and studio transmitter links, among other services. The Enforcement Division reference room keeps station complaint files; Congressional correspondence files; and network correspondence files. The Broadcast Ownership reference room has commercial and noncommercial ownership reports for AM, FM, and TV stations; contract files; and network affiliation agreements. The Fairness/Political Programming Branch maintains correspondence on fairness/political programming complaints. The Cable Television office has various forms relating to registration, ownership exchange, and physical system data; Cable Antenna Relay Services (CARS); cable show cause orders; cable special relief; and cross-ownership files. The Equal Employment Opportunity office has report forms for broadcast and cable; cable company submissions in the certification process; labor force statistics; correspondence to and from cable and broadcast employment units; and cable certification process results. Other miscellaneous Mass Media files are history cards for every station ever licensed, and TV network affiliation information.

Common Carrier Bureau

The Accounting and Audits Division public reference room includes accounts; continuing property records; pension filings; depreciation rates; and contracts between carriers and affiliates. The Domestic Facilities Division public reference room includes files on point-to-point microwave; digital electronic message services; MDS services; space stations; and equipment registration. Mobile Service Division has station files; maps and diagrams; petitions; channel searches; granted cellular station files. The Cellular reference room contains the pending cellular applications and petitions. The Industry Analysis Division public reference room includes reports required by the FCC; annual reports to stockholders; rate of return reports; statistics on common carriers; operating data of telephone and telegraph carriers; local exchange rates; and other reports relating to lineups, pools, and switches access. The Tariff Review reference room includes documents on MDS and AT&T; access tariffs by Bell Operating Companies; independent telephone companies; Western Union; international carriers; specialized common carriers; satellite carriers; microwave carriers; maritime carriers; CATV and wide spectrum facilities; and mobile radio-telephone service. The Formal Complaints and Investigations Branch has formal complaints and pleadings; interlocking directorate reports and applications; pole attachment complaints and certifications; enforcement proceedings such as mergers, acquisitions, and transfer of control. The Informal Complaints and Public Inquiries Branch keeps a record of informal written telephone service related complaints. The Policy and Program Planning Division files include petitions, comments, and replies in non-docketed proceedings; applications for review; and petitions for declaratory rulings and for waivers. The Tariff Legal Branch has NPRMs; petitions against tariffs; petitions for reconsideration; access tariffs; and applications for review. The International Facilities Division includes international files for microwave point-to-point, fixed radiotelephone, fixed radiotelegraph, earth stations, and space stations; private operating agency files; submarine cable landing licenses; and INTELSAT and COMSAT documents.

Private Radio Bureau

The Private Radio Bureau's reference room contains documents on land mobile and GMRS applications; microwave applications; aviation ground applications; marine coast applications; Special Temporary Authority (STA); and transfers of control.

Field Operations Bureau

The Field Operations Bureau's Public Service Division has files on commercial radio operator applications and bulletins concerning radio operator matters.

Office of Engineering and Technology (OET)

Information on authorizations and pending applications regarding type-accepted equipment may be obtained from OET's public access line, 301-725-1072. OET's Laboratory reference room, in Columbia, Maryland, has documents on equipment authorizations; test laboratory information, technical journals, and other technical materials. The OET Technical Library has copies of OET reports; measurement procedures; engineering support material of FCC actions; Spectrum Management Task Force data; technical journals and materials; and historical materials on radio propagation. A Frequency Allocations history room has materials on that subject in chronological order. A Treaty Library includes documents on International Telecommunications Union (ITU) publications, conventions, regulations, administrators' addresses, conferences, reports and recommendations; international frequency assignments; and specialized lists including coast stations, ship stations, and Canadian frequency assignments. The Experimental Radio service has licenses, applications, and background materials pertaining to that service.

Office of the Managing Director

Its Internal Control and Security Office has financial disclosure reports, recusal statements, Freedom of Information Act requests, and ex parte coordination reports.

Office of General Counsel

Its Litigation Division file room has case files of all actions brought against the FCC; and computerized case histories that include lists of pending cases, background histories of cases, status of cases, and deadlines for submissions of briefs.

Finally, for those interested in detailed legal information on communication law and policy as implemented by the FCC, *Pike and Fischer Radio Regulation* is the definitive legal compilation of rule-making proceedings, rules and regulations, federal statutes, and court decisions regarding services under FCC regulatory jurisdiction. These include radio, television, cable, land-mobile, MDS, satellites, cellular radio and, to a lesser degree, common carrier services. The initial volumes date from July 1, 1945. *Pike and Fischer* is divided into specified areas for the different communication services. In addition, there are several volumes that present digests of actions, topically organized. The principal volumes are devoted to full text—with few exceptions—of all cases handled by the FCC, in chronological order. *Pike and Fischer* not only is the standard reference work in communication law offices, but is an excellent research tool, as well, for persons involved professionally with the communication industry and for students and scholars.

Index